THE PARABOLOIDAL REFLECTOR ANTENNA IN RADIO ASTRONOMY AND COMMUNICATION

ASTROPHYSICS AND SPACE SCIENCE LIBRARY

VOLUME 348

THE PARABOLOIDAL REFLECTOR ANTENNA IN RADIO ASTRONOMY AND COMMUNICATION

Theory and Practice

JACOB W.M. BAARS
European Southern Observatory
Swisttal/Bonn, Germany

Springer

Jacob W.M. Baars
European Southern Observatory
Swisttal / Bonn, Germany
jacobbaars@arcor.de

Cover illustration: The APEX Submillimeter Telescope of the Max-Planck-Institut für Radioastronomie on Llano de Chajnantor, Chile. Photo: Arnaud Belloche, Max-Planck-Institut für Radioastronomie, Bonn, Germany.

Frontispiece: Specchi Ustori di Archimede (Archimedé Burning Glasses), Giulio Pangi (attr.) 1599. Courtesy of Galleria degli Uffizi, Polo Museale, Firenze, Italia.

ISBN-10: 1-4899-9708-3
ISBN-13: 978-1-4899-9708-1
ISBN-10: 0-387-69734-9 (eBook)
ISBN-13: 978-0-387-69734-5 (eBook)

Printed on acid-free paper.

9 8 7 6 5 4 3 2 1

springer.com

Dedication

To Dave Heeschen, who enabled me to jump in the water,
Peter Mezger, who taught me to swim,
the memory of Ben Hooghoudt, who showed me how to manage the waves,
Richard Kurz, who saved me from drowning in the "Mexican Gulf".

and

My wife Marja, for love and patiently keeping me afloat throughout.

Preface

The paraboloidal (often called parabolic) reflector is one of the most versatile and widely used antenna types for the transmission and reception of electro-magnetic waves in the microwave and millimeter wavelength domain of the electro-magnetic spectrum. The development of large and highly accurate reflectors has mainly been carried out at radio astronomy observatories. The emergence of satellite communication and deep-space research with satellites necessitated the use of such large reflectors as ground stations for the communication with the satellites and space-probes. Over the years radio astronomers have developed the techniques of calibration of large antennas with radio astronomical methods. These are often the only way to characterise the antenna, because the farfield distance precludes an earth-bound test transmitter and the antennas are too big for nearfield scanning test ranges.

The general theory of the reflector antenna has been presented quite completely in the classic book by Silver (1949) in the MIT Radiation Laboratory Series. Modern approaches of computer-aided analysis and design were discussed by Rusch and Potter (1970). With the current methods of analysis, like the geometrical theory of diffraction and fast algorithms of surface current integration, the analysis of the detailed behaviour of the radiation characteristics can be realised. Nevertheless these methods are laborious and often not suitable for the accurate prediction of the detailed antenna behaviour under non-ideal conditions, as mechanical distortions under gravity, temperature gradients and wind forces. Here a combination of approximate theoretical analysis and measurement of antenna parameters is often the best approach to characterise the antenna.

The techniques developed by radio astronomers for the characterisation of large reflector antennas has not been described comprehensively in the open literature. An early effort in this area is "Radioastronomical Methods of Antenna Measurements" by Kuz'min and Salomonovich (1966). Since then these methods have been further developed and a considerable body of experience is now available, which however has only been sparsely published in readily accessible form. It is the purpose of this book to fill part of this gap. The book is neither a replacement for antenna theory texts like Silver or Rusch and Potter, nor a substitute for books on radio astronomy techniques like Kraus (1966), Rohlfs and Wilson (1996) or Thompson, Moran and Swenson (2001). It is less general than the book "Radiotelescopes" by Christiansen and Högbom (1969). Structural and mechanical aspects of large reflector antennas have been presented in Mar and Leibovich (1969) and by Levy (1996).

Here we are mainly concerned with electromagnetic aspects and concentrate on a discussion of the paraboloidal reflector antenna in a practical approach. The theory is developed with this in mind and considerable attention is given to the treatment of non-ideal situations and the calibration of antenna parameters. While the parabolic reflector is the most used antenna, much of the discussion applies *mutatis mutandis* to spherical and elliptical reflectors, as well as to so-called "shaped" systems.

The general subject of "Instrumentation and Techniques for Radio Astronomy" is well illustrated by the articles in the selected reprints volume of this title, collected and commented by Goldsmith (1988). Because of its usefulness, we shall indicate the presence of particular references in this volume next to their original source in our chapter reference lists as [Gold, pp]. We omit a treatment of polarisation. While control of cross-polarisation is certainly of importance in communication systems and in a limited, but important part of radio astronomy observations, its full discussion is beyond the aims set for this text. A complete treatment of polarisation can be found in a book by Tinbergen (1996). Basic and practical aspects of polarisation in radio interferometry are described by Morris et al.(1964) and Weiler (1973). An original matrix treatment of radio interferometric polarimetry is presented in a series of three papers by Hamaker, Bregman and Sault (1996).

All calculations and results in the form of tables and figures have been made with the aid of the software package *Mathematica* from Wolfram Research (Wolfram, 1999). The *Mathematica* instructions and expressions can be used directly by the reader with access to *Mathematica* for the implementation of his own input data. In order not to break the flow of the text, the routines are assembled at the end of each chapter. In the text they are identified in blue print as [Mat.x.y]. Actually, the entire book has been written as a *Mathematica* Notebook using the excellent editorial capabilities of the program. It is hoped that the availability of the *Mathematica* routines will contribute to the usefulness of the book in daily use. The routines are being made available for download on the Springer Website. The reference lists are not exhaustive. We provide references of a historical nature, original work used in the text and selected references for further study of details.

We aim to address the needs of observational radio astronomers and microwave communication engineers. The book should be of use to all who are involved in the design, operation and calibration of large antennas, like ground station managers and engineers, practicing radio astronomers and graduate students in radio astronomy and communication technology.

Acknowledgement.

The material assembled in this book has been part of my work for more than 40 years. It is clear that over such a time span I have been influenced by and have used contributions from numerous colleagues to whom I owe my debt. They are too numerous to list, but I have endeavoured to give them proper credit in the references. A few of them must be mentioned because of their decisive influence on my work and interests. Peter Mezger introduced me to the subject of antenna calibration and guided me through my early efforts. John Findlay was a great source of inspiration in the early years. The long and close collaboration with Albert Greve and Jeffrey Mangum is reflected in the text at many places. I also thank both of them for a valuable scrutiny of the draft text.

Swisttal / Bonn, October 2006 Jacob W.M. Baars

Bibliography

Christiansen, W.N. and J.A. Högbom, *Radiotelescopes*, Cambridge, University Press, 1969.

Goldsmith, P.F. (Ed.), *Instrumentation and Techniques for Radio Astronomy*, New York, IEEE Press, 1988. [Referred to as "Gold, pp" for individual articles.]

Hamaker, J.P., J.D. Bregman and R.J. Sault, Understanding radio polarimetry, *Astronomy and Astrophysics Sup.* **117**, 137-147, 149-159, 161-165 (3 papers), 1996.

Kraus, J.D., *Radio Astronomy,* New York, McGraw-Hill, 1966; 2nd Ed., Powel OH, Cygnus-Quasar Books, 1986.

Kuz'min, A.D. and A.E. Salomonovich, *Radioastronomical Methods of Antenna Measurements*, New York, Academic Press, 1966.

Levy, R., *Structural Engineering of Microwave Antennas*, New York, IEEE Press, 1996.

Mar, J.W. and H. Liebowitz (Eds.), *Structures Technology for large Radio and Radar Telescope Systems*, Cambridge MA, MIT Press, 1969.

Morris, D., V. Radhakrishnan and G.A. Seielstad, On the Measurement of Polarization Distribution over Radio Sources, *Astrophys. J.* **139**, 551-559, 1964.

Rohlfs, K. and T.L. Wilson, *Tools of Radio Astronomy*, 2nd Ed., Berlin, Springer, 1996.

Rusch, W.V.T. and P.D. Potter, *Analysis of Reflector Antennas*, New York, Academic Press, 1970.

Silver, S., *Microwave Antenna Theory and Design,* MIT Rad. Lab Series **12**, New York, McGraw-Hill, 1949.

Tinbergen, J., *Astronomical Polarimetry*, Cambridge, University Press, 1996.

Thompson, A.R., J.M. Moran and G.W. Swenson, *Interferometry and Synthesis in Radio Astronomy*, New York, Wiley, 2001.

Weiler, K.W., The Synthesis Radio Telescope at Westerbork, Methods of Polarization Measurements, *Astron. Astrophys.* **26**, 403-407, 1973.

Wolfram, S., *The Mathematica Book*, 4th Ed., Wolfram Media/ Cambridge University Press, 1999.

Contents

List of Figures and Tables

■ Chapter 1

■ Chapter 2

■ Chapter 3

■ Chapter 4

■ Chapter 5

■ Chapter 6

■ Chapter 7

■ List of Tables

Definition of Symbols

A aperture area
B brightness (W m^{-2} Hz^{-1})
C atmospheric structure function constant
D directivity, structure function
E electric field
F aperture field distribution
G gain
H magnetic field
I intensity
J jansky (10^{-26} W m^{-2} Hz^{-1})
K beam smearing correction factor
L atmospheric pathlength
N atmospheric refractivity
P power
R distance to field point
 R_f farfield distance
 R_r Rayleigh distance
S flux density (unit is jansky)
T temperature
 T_A antenna temperature
 T_B brightness temperature
U Lommel function

a radial aperture coordinate
b beamwidth coefficient
c speed of light, correlation length
d diameter of antenna
e Cassegrain eccentricity
f amplitude pattern, focal length
g antenna power pattern
h Planck's constant
i imaginary unit
j current density
k Boltzmann's constant, wave number
l atmospheric scale length
m Cassegrain magnification
n refractive index
p number beamwidths off-axis
r distance
t time
u,v,w direction cosines
x,y,z cartesian coordinates

α absorption coefficient
γ help variable (Ch.4)
ϵ surface error, permittivity
η efficiency
 η_A aperture efficiency
 η_B beam efficiency
λ wavelength
ρ distance to aperture point
τ illumination taper
ϕ azimuthal angle of field point
χ azimuthal angle in aperture

β phase error
δ small deviation, feed offset
θ polar angle of field point
μ permeability
ν frequency
σ random phase error
ψ polar angle to aperture
ω circular frequency

Γ help variable (Ch.4)
Θ angle
Σ sum
Φ reflector opening angle
Ω solid angle

Δ deviation, path error
Λ lambda function
Ξ $= 4\, f\, /\, d$ (Ch.4)
Ψ subreflector opening angle

BDF beam deviation factor

HPBW half-power beamwidth

1. Introduction and historical development

■ 1.1. Some history of the parabolic reflector antenna

In highschool I was taught that the great Greek scientist Archimedes (287 - 212 BC) was instrumental in the defence of his city Syracuse on Sicily against the Roman fleet and army of Marcellus during the Punian War. Among the defensive devices, developed by him, he used parabolic mirrors to concentrate the reflected light from the Sun to set the Roman ships afire. The frontispiece picture on page vi of this book represents this feat. Although this story is now considered by historians to be very likely not true, it is fair to say that Archimedes could have done it. He had after all studied the geometrical figures of *conic sections* and knew about the focussing characteristics of such curves. As such, it would have been one of the earliest examples of applied physics based on pure mathematical knowledge. Note that for his goal to be successful, he needed to construct what is now called an offset reflector, a typical example being the ubiquitous TV-satellite dish. From the definition of the parabola it is immediately clear that a bundle of parallel light rays impinging onto a reflector in the form of a paraboloid of revolution along its symmetry axis will be concentrated towards the focal point of the paraboloid. This simple characteristic has made the parabolic reflector the most widely used device for astronomical telescopes, both in the optical and radio regime, and more recently for transmitting and receiving antennas in microwave communication technology, including satellite communication, as well as the concentration of solar radiation as commercial energy source in solar power stations.

In this book we want to explore the characteristics of the paraboloidal reflector, and other types related to it, with special emphasis on its use as a radio telescope or communication antenna. The geometry and electromagnetic theory are treated first, followed by detailed discussions of the application and calibration of large antennas for radio astronomy. The approach of the treatment is practical. Many formulae, curves and tables will be derived with application in mind, rather than aiming at the highest level of mathematical rigor. The vehicle chosen for the calculations and most of the figures is the software package *Mathematica*. The reader may copy the formulae into his own copy of *Mathematica* and vary the parameters according to his own requirements.

The theory of electromagnetic (EM) waves was developed by James Clerk Maxwell (1865). He showed that light can be described by this theory and predicted the existence of EM-waves of other wavelengths. The experimental demonstration of EM-waves with wavelengths of what we now call radio waves was achieved in 1888 by Heinrich Hertz (1888). In his experimental setup he used cylindrical paraboloids of 2 m length, aperture width 1.2 m, depth 0.7 m to concentrate the waves with a wavelength of about 66 cm onto the wire-antenna along the focal line of the reflector. He concluded that "radio waves" ("Strahlen elektrischer Kraft") are identical to light rays with large wavelength. His experiments formed the beginning of the enormous development of radio in the twentieth century. It also caused several people to consider the possibility that radio waves may be emitted by the Sun. In the period 1897-1900 Sir Oliver Lodge described his plans for detecting radio waves from the Sun before the Royal Academy and in his book "Signaling across Space without Wires", 3rd Edition (1900). Other proposals and experiments came from Thomas A. Edison in the USA as well as from France and Germany. They were all unsuccessful.

Fig. 1.1. Reber's 10 m antenna, reconstructed at the National Radio Astronomy Observatory, Green Bank, WV, USA. The original dish did not have the azimuth movement; it was a transit instrument. (NRAO/AUI/NSF)

In 1932 Karl Guthe Jansky (1933) discovered radio radiation from space, while studying the interference from thunder storms on short-wave communication with a directional wire antenna at a wavelength of about 20 m. His proposal to build a large parabolic reflector to systematically study the "cosmic radio waves" was not successful, but Grote Reber, a radio engineer, became intrigued by the prospect of radio astronomy and in 1937 designed and built a 10 m diameter paraboloid in his backyard (Fig.1.1). This formed the beginning of radio astronomy and Reber (1940) published the first map of radio radiation from our Galaxy in the Astrophysical Journal, as well as the Proceedings of the IRE (Institute of Radio Engineers, now the IEEE).

The development of military radar, both in Germany and the allied countries during the second world war resulted in the construction of radar antennas, many in the form of paraboloidal reflectors. The German "Würzburg Riese" (giant of Würzburg) was a 7.5 m diameter reflector with a mesh surface. It was adapted after WW2 by Dutch, British, French and Scandinavian radio astronomers as their first radio telescopes. Most notably, the detection of the spectral line of neutral hydrogen at 21 cm wavelength in Holland was made with a Würzburg antenna (Fig.1.2), as was the observation of the first complete hydrogen line survey of the Galaxy by van de Hulst, Muller and Oort (1954). In an odd turn of history, the antenna has been returned to Germany for exhibition at the Deutsches Museum in Munich.

Fig.1.2. A 7.5 m diameter "Würzburg" antenna, mounted equatorially (foreground) at the Dwingeloo Radio Observatory, Netherlands. The 25 m telescope is in the background, Both antennas were used as an interferometer in an experiment described in Ch.6.2. (Henk Snijder, ASTRON)

In the nineteen-fifties larger paraboloidal radio telescopes appeared on the scene; the 50-ft NRL dish on the Potomac River in 1951 had a high surface accuracy of 1 mm rms and enabled observations at wavelengths as short as 3 cm (Hagen, 1954). The Dutch 25-m telescope in Dwingeloo, completed in 1956 (Fig.1.2), was for some time the largest fully steerable radio telescope in the world until it was overtaken by the gigantic 76-m diameter telescope in Jodrell Bank, UK in 1957 (Fig.1.3) (Lovell,

1957). The design and construction of these ever larger instruments has mainly been performed at radio observatories in collaboration with structural and mechanical engineers from industry.

Fig.1.3. The 76 m diameter Lovell radio telescope at Jodrell Bank Observatory, England. The antenna has undergone two major upgrades, including a new, more accurate surface with a larger focal ratio. The central post to support the feed has been maintained. (I. Morison,Jodrell Bank Observatory)

The era of space communication became a reality soon after the launching of Sputnik in 1957. To use satellites for this purpose it was necessary to build powerful ground stations for both the uplink and downlink connections. Thus parallel to the large increase in radio telescopes between 1960 and 1980, the number of 25-35 m diameter satellite ground stations is even larger. The tracking stations for the deep-space efforts of the NASA and ESA (Fig.1.4) are equipped with antennas as large as 70 m and are now made suitable for operation in the 30 GHz band (Imbriale, 2003). Some of these have also been used for radio astronomy, notably as stations in the global and geodetic VLBI (very long baseline interferometry) systems.

The signals received from cosmic objects are extremely weak and generally broad-band. Thus the larger the reflector surface area, the better the astronomical result. Moreover, observations normally need to be done over widely separated frequencies. Considering the very poor angular resolution of radio telescopes (typically many arcminutes in the early days of its development compared to about 1 arcsecond in the optical domain), it was also obvious that there emerged a push towards higher frequencies with a concomitant higher angular resolution. However, working at higher frequencies requires a reflector surface of higher accuracy of shape in order not to scatter the reflected radiation away from the focal region.

Fig.1.4. Example of a modern deep-space ground station. This 35 m diameter ESA antenna is located in Western Australia and was designed and built by Vertex Antennentechnik of Germany. (ESA/ESOC)

The development of radio telescopes thus resulted in larger and simultaneously more accurate reflectors and highly stable and accurate mounting structures and pointing control systems. The first telescope of this genre was the 140-ft (43 m) diameter radio telescope of the National Radio Astronomy Observatory in Green Bank, West Virginia. It was conceived in the late fifties with a goal of reaching 2 cm as a shortest wavelength. In those days there were relatively few radio astronomers in the USA and, in particular, those with an engineering background were rare. Conservatism led to the decision to build the telescope with an equatorial mounting, because there was doubt about analog coordinate converters and the capabilities of computers to do the necessary computations of coordinate transformation. The result was a protracted design and construction process which ended in 1965 with the dedication of what became one of the most productive radio telescopes ever built (Fig.1.5). The interesting story of the "140-ft" has been told, albeit with errors, by Malphrus (1996).

Fig.1.5. The NRAO !40-ft telescope. The polar axis runs on hydrostatic bearings. The reflector structure and panels are made from aluminium. The original prime focus geometry was converted to Cassegrain and the receiver box in the secondary focus protrudes from the vertex of the main reflector. (NRAO/AUI/NSF)

In 1972 the 100-m Effelsberg radio telescope of the Max-Planck-Institut für Radioastronomie came into operation (Hachenberg et al., 1973). This telescope (described in Ch.7), and the slightly larger Green Bank Telescope (Fig.1.6) of the National Radio Astronomy Observatory (put into use in 2001, see Jewell and Prestage, 2004 and Lockman, 1998) reach a shortest wavelength of 3 mm with acceptable performance.

Special millimeter and submillimeter telescopes have been built since the seventies. For these to be effective observing tools, it is necessary to locate them on a high and dry site, so as to minimise the absorption of the short wavelength radiation by the troposphere. A highly successful millimeter telescope is the 30-m antenna of IRAM in southern Spain (see Ch.7), which operates to a shortest wavelength of 0.8 mm (Baars et al., 1987). Special submillimeter telescopes of 10 - 15 m diameter reach surface accuracies of 12 - 25 μm and can operate to frequencies as high as 1 THz. We discuss the Heinrich Hertz Telescope in Chapter 7 and show here pictures of the James Clerk Maxwell Telescope (JCMT) (Fig.1.7) and the Caltech Submillimeter Observatory (CSO) (Fig.1.8), both sited at 4000 m altitude on Hawaii. Both are placed in a protective dome, which opens fully for the 10 m diameter CSO, while the 15 m diameter JCMT normally operates behind a goretex screen, which transmits most of the submillimeter wavelength radiation but provides protection against wind and fast temperature variations.

Fig.1.6. The Green Bank Telescope (GBT) of NRAO. This "clear aperture" antenna is an off-set piece of a paraboloid of 100 m diameter and provides a very clean beam with low far-sidelobes. Like the Effelsberg antenna it uses a Gregorian focus arrangement with elliptical secondary reflector. Prime focus operation is used for long wavelengths. (NRAO/AUI/NSF)

The 12 m diameter antennas of the ALMA Project, described in Chapter 7, however will operate exposed to the harsh environment at 5000 altitude in the Atacama Desert of Northern Chile. The Max-Planck-Institut für Radioastronomie of Germany has adapted one of the ALMA prototype antennas for single dish work and located it at the ALMA site. This antenna is called ALMA Pathfinder Experiment (APEX) and is shown in Fig.1.9. It has a surface accuracy of 17 μm and a pointing stability of better than 1 arcsecond (Güsten et al., 2006).

For the characterisation and calibration of these large reflector antennas, techniques have been developed by radio-astronomers, which use the existence of a number of strong cosmic radio sources. These methods have been adopted by operators of satellite ground stations. The sheer size of these antennas, counted in wavelengths of

Fig.1.7. The 15 m diameter JCMT on Hawaii. The protective goretex screen has been rolled up, exposing the reflector. (Robin Phillips, courtesy JCMT, Mauna Kea Observatory, Hawaii)

Fig.1.8. The 10 m diameter CSO submillimeter wavelength antenna on Hawaii. (Submillimeter Observatory, California Institute of Technology)

Fig.1.9. The 12 m diameter APEX antenna on the ALMA site in Chile, derived from one of the ALMA prototype designs. To accommodate large multi-feed receivers, the antenna uses a Nasmyth focus with two large equipment rooms at either side of the elevation bearings. (ESO / Max-Planck-Institut für Radioastronomie)

the received radiation, makes it impossible to measure the far-field characteristics on a terrestrial measurement range. The techniques of antenna measurement with the aid of cosmic sources have not been comprehensively described in the open literature. It is the aim of this book to fill this gap. We limit ourselves to the case of a single antenna. Problems associated with the operation and calibration of interferometric arrays are not treated here. For this we refer the reader to the excellent book by Thompson, Moran and Swenson (2001).

In the last chapter we describe original design features of a number of important radio telescopes to illustrate the state of the art.

■ 1.2. Measuring antenna parameters with cosmic radio sources

The measurement of the far-field characteristics of large antennas, of a size larger than a thousand wavelengths, say, is difficult for the simple reason that the required distance to the test transmitter becomes prohibitively large. For accurate measurements one wants to avoid corrections for the finite distance to the test source, which consequently has to be at a distance larger than the **far-field distance**, defined as

$$R_f = 2d^2 / \lambda, \tag{1.1}$$

where d is the diameter of the antenna aperture and λ the wavelength. For an antenna with a diameter of 12 m, operating at a wavelength of 3 mm, we find $R_f = 96\,\text{km}$!

Because the historical development of large reflector antennas has been led by radio astronomers, it is not surprising that radio astronomers have turned to extraterrestrial radio sources as test transmitters for the characterisation of their radio telescopes. A cosmic source exhibits a number of clear advantages. It is definitely in the far-field and it has a fixed celestial position, which means that it describes a well determined diurnal path on the sky as seen from the telescope. This enables immediately the measurement of any antenna parameter as function of the elevation angle. The obvious application here is the determination of the pointing model of the antenna, for which a set of radio sources with known positions and distributed over the celestial sphere is required. If we want to use these sources for the characterisation of the antenna beam parameters, knowledge of the absolute intensity of the source at the measurement frequency is necessary. Here we meet our first significant obstacle: to determine the absolute flux density of the source, we need to measure it with an antenna of known gain. However, absolute calibration of the gain can only be done for small antennas (the far-field restriction) or by calculation on simple geometries, like horns and dipoles. Thus, a signal to noise ratio sufficient for accurate results can only be achieved with very strong radio sources. As an example, we consider the "Little Big Horn" in Green Bank (Findlay et al., 1965). The LBH has an aperture of about 5x4 m^2 and operates at a frequency of 1400 MHz, where the calculated gain is 34.27 dB, equivalent to an effective absorption area of 9.76 m^2 (Fig.1.10).The flux density of Cassiopeia A, the strongest cm-wavelength source in the sky (ignoring the Sun because of its large angular size of 0.5 degree), is about 2400 Jy (Jansky, 1 Jy = 10^{-26} W m^{-2} Hz^{-1}), providing an antenna temperature at the output of the LBH of approximately 8.5 K. For a signal-to-noise ratio, SNR > 100 we need to restrict the receiver noise to < 0.1 K, which leads to a required receiver noise temperature (using 5 MHz bandwidth and 10 sec integration time) of 600 K. These were indeed typical values at the time of the LBH measurements. Note that the LBH was by far the biggest horn ever used for this type of work.

There was however no other way to reliably establish antenna parameters of large telescopes and in the years 1955-1975 a number of workers built absolutely calibrated small antennas and established the absolute spectra of the strongest radio sources over a good part of the electromagnetic spectrum, roughly from 100 MHz to 15 GHz. An analysis of these observations was carried out by Baars, Mezger and Wendker (1965). A more definitive summary appeared in 1977 (Baars, Genzel, Pauliny-Toth and Witzel, 1977), in which a set of secondary standards, consisting of weaker but point-like sources was established which became the *de facto* flux density scale for cm-wavelength radio astronomy. The very strong sources (Cas A, Cyg A and Tau A) all have an angular extent which is significant with respect to the beamwidth of the larger short wavelength telescopes, which causes the need for corrections to the measurements. Because of their point-like size, the secondary point sources on the other hand can also be used as flux calibrators for interferometric arrays.

Fig.1.10. The 25 m long "Little Big Horn" at NRAO, Green Bank, used for absolute flux density measurements of Cas A at 1400 MHz over many years. The rotation of the earth carried the source through the beam once a day. (NRAO/AUI/NSF)

The strongest sources mentioned above are all non-thermal radiators and their flux density becomes small at high frequencies in the millimeter wavelength range. Also, with an angular size of several arcminutes, they become significantly larger than the average beamwidth of the newer, large mm-telescopes, which typically have a beamwidth of less than one arcminute. Fortunately, there is a small group of objects which are well suited for calibration purposes at the high frequencies, namely the planets and some asteroids (and to a lesser degree the Moon). There are however significant complications in the use of the planets. While distant enough to obey the far-field criterion, they nevertheless change significantly in distance to the Earth, which causes a variable angular diameter and flux density. Fortunately their orbits are well known and ephemerides readily available to overcome this handicap. Considerable effort has been devoted over the last 30 years to establish absolute brightness temperatures of the planets (Ulich et al, 1980). With care, the mm-wavelength brightness temperature of most of the planets can now be predicted with an absolute error of 5 to 10 percent over the entire mm- and submm- wavelength region. This is less accurate than the strong sources in the cm-range and improvements in the accuracy would be necessary to achieve, for instance, the calibration goal for ALMA of five percent in flux density.

A considerable part of this book is concerned with the methods for antenna characterisation with the aid of cosmic sources. In addition to the determination of the gain, we shall want to measure the beam shape and near sidelobe level. In general, the interaction between the beam with its angular structure and the radio source, which normally will have a finite angular extent and perhaps an irregular brightness structure needs to concern us in order to extract quantities intrinsic to the source under study. We shall see how observations of sources of varying angular size can provide valuable information about the characteristics of the antenna beam. Also, measurements of the gain over a range of frequencies can deliver an estimate of the accuracy of the shape of the reflector. Thus, while it is essential to know the parameters of the antenna to properly analyse the astronomical observations, it is the observation of some of these sources which provides us with knowledge of the beam parameters.

In Chapter 2 we present the geometry of the paraboloidal reflector antenna. Both the general geometrical relations and the geometry of aberrations (out of focus situations) will be treated. In Chapter 3 the basic theory of the paraboloidal antenna is summarised. The differences between Fraunhofer and Fresnel diffraction are described. We then turn to more practical applications of antenna theory in Chapter 4, where the major formulae for the characterisation of large reflector antennas are derived. This chapter contains a wealth of information in the form of graphs, tables and formulae, directly usable for the analysis of antenna measurements. In Chapter 5 we discuss the interaction between the antenna and the source. Here we present the methods and formulae for the measurement of antenna parameters with the aid of cosmic radio sources. Data on radio sources suitable for antenna measurements are summarised. In the sixth chapter we deal with some special aspects. For instance the holographic method of antenna surface measurements is treated there in quite a detail. Also attention is given to the influence of the earth's atmosphere and special methods of observation are described, among those the suppression of atmospheric influences and methods of calibration. In the final chapter we present short descriptions of the design and performance features of a number of important radio telescopes, highlighting their characteristics as measured with the methods described here.

References

Baars, J.W.M., P.G. Mezger and H. Wendker, Spectra of the strongest non-thermal Radio Sources in the centimeter wavelength range. *Astrophys. J.* **142**, 122-134, 1965.

Baars, J.W.M., R. Genzel, I.I.K. Pauliny-Toth and A. Witzel, The Absolute Spectrum of Cas A; An accurate flux density scale and a set of secondary calibrators. *Astron. Astrophys.* **61**, 99-106, 1977 [Gold, 401].

Baars, J.W.M., B.G. Hooghoudt, P.G. Mezger and M.J. de Jonge, The IRAM 30-m millimeter radio telescope on Pico Veleta, Spain. *Astron. Astrophys.* **175**, 319-326, 1987 [Gold, 56].

Findlay, J.W., H. Hvatum and W. B., Waltman, An absolute flux density measurement of Cassiopeia A at 1440 MHz. *Asrophys. J.* **141**, 873-884, 1965.

Güsten, R., L.Å. Nyman, P. Schilke, K. Menten, C. Cesarsky and R. Booth, The Atacama Pathfinder EXperiment (APEX) - a new submillimeter facility for southern skies, Astron. Astrophys. **454**, L13-16, 2006.

Hachenberg. O., B.H. Grahl and R. Wielebinski, The 100-meter Radio Telescope at Effelsberg. *Proc. IEEE* **61**, 1288-1295, 1973 [Gold, 32].

Hagen, J.P., "NRL - 50 ft machined aluminium dish", *J. Geophys. Res.* **59** ,183, 1954.

Hertz, H., Über elektrodynamische Wellen im Luftraume und deren Reflexion, *Wiedemanns Annalen* **34**, 610, 1888.

Hertz, H., Über Strahlen elektrischer Kraft, *Wiedemanns Annalen* **36**, 769, 1888.

Hulst, H.C. van de, C.A. Muller and J.H. Oort, The spiral structure of the outer part of the Galactic system derived from the hydrogen emission at 21 cm wave length, *Bull. Astron. Institutes of the Netherlands* **12**, 117-149, 1956.

Imbriale, W.A., *Large Antennas of the Deep Space Network*, New York, John Wiley, 2003.

Jansky, K.G., Radio waves from outside the solar system. *Nature* **132**, 66, 1933.

Jewell, P.R. and R.M. Prestage, The Green Bank Telescope, *Proc SPIE* **5489**, 312-323, 2004.

Lockman, F.J., The Green Bank Telesccope: An Overview, *Proc. SPIE* **3357**, 656-665, 1998.

Lodge, O., *Signaling across Space without Wires*, 3rd Edition, 1900.

Lovell, A.C.B., The Jodrell Bank Radio Telescope, *Nature* **180**, 60-62, 1957 [Gold, 21].

Malphrus, B.K., *The History of Radio Astronomy and the National Radio Astronomy Observatory: Evolution Toward Big Science*, Melbourne, FL, Krieger Publ. Co.,1996.

Maxwell, J. Clerk, A Dynamical Theory of the Electromagnetic Field, *Royal Society Transactions* **65,** 1865.

Reber, G., Cosmic Static. *Proc. IRE* **28**, 68-70, 1940.

Thompson, A.R., J.M. Moran and G.W. Swenson, *Interferometry and Synthesis in Radio Astronomy*, 2nd Ed., New York, John Wiley, 2001.

Ulich, B.L., J.H. Davis, P.J. Rhodes and J.M. Hollis, Absolute Brightness Temperature Measurements at 3.5-mm Wavelength, *IEEE Trans. Antennas Propagat.* **AP-28**, 367-377, 1980 [Gold, 413].

2. Geometry of reflector antennas

In this chapter we deal with a description of the geometry of a paraboloid of revolution. We shall collect the geometrical relationships, which we need for the description of the electromagnetic radiation characteristics of the paraboloidal reflector antenna. Because many radio telescopes and communication antennas actually employ the Cassegrain or Gregorian layout, we include the formulas for those systems too. These are dual reflector systems, where a relatively small secondary reflector, placed near the focus of the primary paraboloid concentrates the received radiation in a secondary focus, located near or behind the vertex of the primary reflector. The Cassegrain employs a hyperboloidal secondary, while the Gregorian system uses an elliptical secondary reflector. One of the foci of these dual-focus conic sections coincides with the focus of the paraboloid, while the other provides the secondary focus at a convenient location. The great advantage from an operational viewpoint is the possibility to locate bulky receiving equipment behind the primary reflector. As we shall see later, there are also significant electro-magnetic advantages of the dual-reflector varieties.

In the second half of this chapter we discuss the aspects of imperfections in the geometry of the reflector system. In particular, the cases of "defocus" are described, where the detector element is displaced from the true focal point. This leads to the so-called *aberrations* in the optical system. We treat those by calculating the path-length differences over the reflector. These are then introduced as phase errors in the electromagnetic analysis of the antenna in a later chapter.

2.1. Geometrical relations of the dual reflector system

The geometrical definition of a parabola is illustrated in Fig. 2.1, where we limit ourselves to the two-dimensional case. Consider a coordinate system (x, z) and choose a point F on the z axis, at a distance f from the origin. Draw a line perpendicular to the x-axis through F and choose a point Q at coordinate (x, z). The definition of the parabola is the locus of points P, where the sum of the distance from point P to F and the distance from P to Q, with PQ parallel to the z-axis, is constant. Now assign Q to F. Then P moves to the origin O and we have "QP" +"PF" = "FO" + "OF" = 2 f. For arbitrary value x of Q on the line through F we have

$$PQ = f - z_P \text{ and } FP^2 = FQ^2 + PQ^2 = x_P{}^2 + (f - z_P)^2 ,$$

where z_p is the z-coordinate of P.

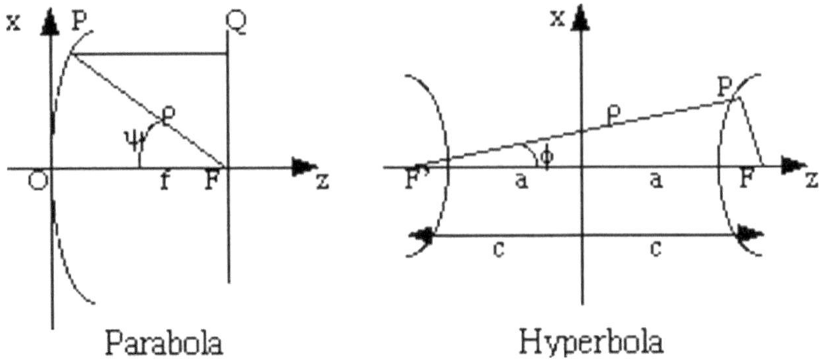

Fig. 2.1. The geometry of a parabola (left) and hyperbola (right).

The definition of the parabola now results in the following equation

$$\mathrm{FP} + \mathrm{PQ} = \sqrt{x_p{}^2 + (f - z_p)^2} + f - z_p = 2f,$$

from which it is easy to derive the defining equation for the parabola in cartesian coordinates as

$$x^2 = 4fz.$$

The definition in spherical coordinates (ρ, ψ) is even simpler:

$$\mathrm{FP} + \mathrm{PQ} = \rho + \rho\cos\psi = 2\,\mathrm{f},$$

from which follows

$$\rho = 2f / (1 + \cos\psi).$$

From the geometrical definition it is easy to see the physical significance of the parabola. Let a bundle of light rays travel parallel to the z-axis. Upon reflection at the parabola, each ray will arrive at point F along equal path length and the intensities of the rays will be added there. The point F is the focus of the parabolic mirror. In terms of wave fronts we can say that a plane wave, traveling along the z- axis is transferred upon reflection at the parabola into a spherical wave converging towards the focal point F and adding the field contributions in phase. Conversely, a source of spherical waves in F will cause a plane wave traveling along the z-axis after reflection at the parabola.

In the Cassegrain antenna a hyperbolic secondary reflector is used to transfer the spherical wave front traveling from the parabola into another spherical wave front towards a secondary focus. The definition of a hyperbola is the locus of points, where the difference between the distances to two fixed points is constant. From the geometry of Figure 2.1 we have PF' - PF = constant. If we put P on the z-axis, it is immediately clear that the constant = 2 a. In the general case of arbitrary P we have

$$\sqrt{(z+c)^2 + x^2} - \sqrt{(z-c)^2 + x^2} = 2a ,$$

from which we can derive the defining formula for the hyperbola

$$\frac{z^2}{a^2} - \frac{x^2}{c^2 - a^2} = 1.$$

It is easy to see that the pathlength of the rays to the secondary focus F_s in the Cassegrain system is constant. In the parabola the path from the plane through the focal point via the reflector to the focus is 2 f. It is intercepted by the hyperbola at point P and then directed towards F_s. This subtracts PF_p from the path and adds PF_s; but PF_s - PF_p = 2 a. Thus the path to F_s is constant, which defines the secondary focus.

An occasionally applied variation of the dual-reflector system is the Gregorian antenna, in which the secondary reflector is a part of an ellipse. An ellipse is defined as the locus of points, where the sum of the distance to two fixed points is constant. The reader can verify that the defining formula for the elliptical reflector in cartesian coordinates is

$$\frac{z^2}{a^2} + \frac{x^2}{c^2 - a^2} = 1 .$$

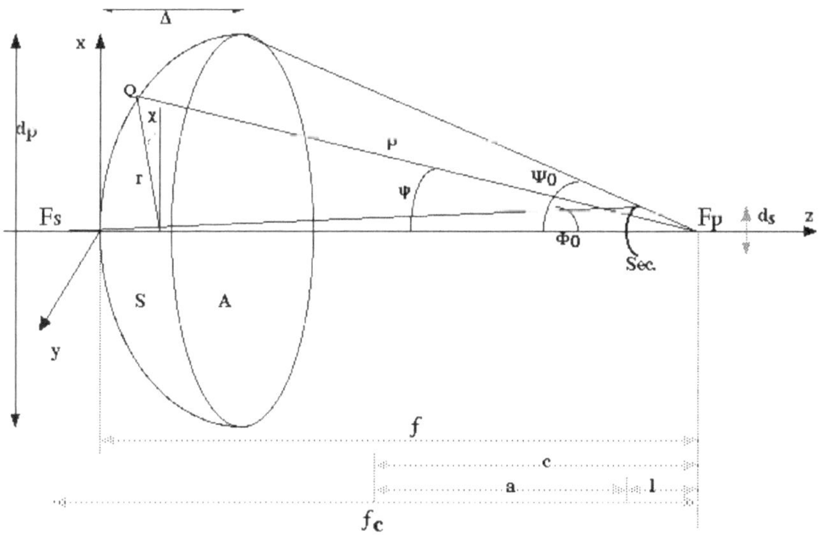

Fig. 2.2. Geometry of the Cassegrain reflector antenna

We turn now to the discussion of the three dimensional paraboloid of revolution. The geometry is illustrated in Fig. 2.2. The finite paraboloid of revolution with surface S, diameter d_P and aperture area A has its focal point F_p at the origin of the cartesian coordinate system and its axis along the z-axis. We denote the focal length by f. A point Q on the surface of the paraboloid (source point) is determined by its spherical coordinates (ρ, ψ, χ). Later we shall also need a point of observation ("field point") P, which we denote by spherical coordinates (R, θ, ϕ). The relations between the unit vectors in these spherical coordinates and the right-handed cartesian system (x, y, z), centred at F_p can be written as

$$\{\, \boldsymbol{i}(R,\ \theta,\ \phi)\,\} = A\ \{\, \boldsymbol{i}(x,\ y,\ z)\,\} \tag{2.1}$$

and

$$\{\, \boldsymbol{i}(\rho,\ \psi,\ \chi)\,\} = B\{\, \boldsymbol{i}(x,\ y,\ z)\,\}, \tag{2.2}$$

where the matrix A reads

$$\mathrm{A} = \begin{pmatrix} \sin\theta\cos\phi & \sin\theta\sin\phi & \cos\theta \\ \cos\theta\cos\phi & \cos\theta\sin\phi & -\sin\theta \\ -\sin\phi & \cos\phi & 0 \end{pmatrix} \tag{2.3}$$

The matrix B has the same form with θ and ϕ replaced by ψ and χ, respectively. Conversely, the unit vectors in the system (x, y, z) are obtained by multiplication of the spherical unit vectors by the transposed matrix $\tilde{\mathrm{A}}$ or $\tilde{B}$. In the notation of *Mathematica* these relations are written as follows:

```
A = {{Sin[θ] Cos[φ], Sin[θ] Sin[φ], Cos[θ]},
    {Cos[θ] Cos[φ], Cos[θ] Sin[φ], -Sin[θ]},
    {-Sin[φ], Cos[φ], 0}};
i = {x, y, z};
A.i
Transpose[A] // MatrixForm
```

```
{z Cos[θ] + x Cos[φ] Sin[θ] + y Sin[θ] Sin[φ],
 x Cos[θ] Cos[φ] - z Sin[θ] + y Cos[θ] Sin[φ],
 y Cos[φ] - x Sin[φ]}
```

```
( Cos[φ] Sin[θ]   Cos[θ] Cos[φ]   -Sin[φ] )
( Sin[θ] Sin[φ]   Cos[θ] Sin[φ]   Cos[φ]  )
( Cos[θ]          -Sin[θ]         0       )
```

The results (output) are the product of Eq. (2.1) and the transposed matrix of matrix A, respectively.

The equation of the paraboloid is

a) in spherical coordinates $\rho = 2f/(1+\cos\psi) = f\sec^2(\psi/2)$ (2.4)

b) in cartesian coordinates $r^2 = x^2 + y^2 = 4f(f+z)$ (2.5)

c) in parabolic coordinates $\rho = \frac{1}{2}(u^2 + v^2)$, (2.6)

where the paraboloidal coordinates (u, v, χ) are connected to the cartesian ones by

$$x = uv\cos\chi,\quad y = uv\sin\chi,\quad z = \tfrac{1}{2}(u^2 - v^2),$$

and χ is the azimuthal angle about the z-axis.

In this coordinate system the paraboloid's surface is described by u = constant (or v = constant). At the vertex V we have v = 0, hence $z = \frac{1}{2}u^2 = f$ and thus $u = \sqrt{2f}$. Also d$\rho = v\,dv$ for u = constant. The infinitesimal surface element is

$$dS = \sqrt{u^2 + v^2}\; uv\,dv\,d\chi = 2\sqrt{f\rho}\,d\rho\,d\chi = 2\cos(\tfrac{\psi}{2})\,\rho\,d\rho\,d\chi, \qquad (2.7)$$

where we have used Eq. (2.4). On eliminating ρ we obtain for the surface element

$$dS = 2f^2\frac{\sin(\psi/2)}{\cos^4(\psi/2)}\,d\psi\,d\chi \qquad (2.8)$$

and the surface of the paraboloid with aperture half-angle Ψ_0 follows as the integral of Eq. (2.8)

$$S = \frac{8\pi}{3}f^2\left[\sec^3\left(\tfrac{\Psi_0}{2}\right) - 1\right] = \frac{8\pi}{3}f^2\left[\left\{1 + \left(\tfrac{d_P}{4f}\right)^2\right\}^{\frac{3}{2}} - 1\right]. \qquad (2.9)$$

Here we have used the alternative expression

$$\tan(\psi/2) = r/2f,\quad (r^2 = x^2 + y^2), \qquad (2.10a)$$

which is easy to derive from Eq. (2.4), considering that $\sin\psi = r/\rho$. From this also follows

$$\tan\left(\frac{\Psi_0}{2}\right) = \frac{d_P}{4f}. \tag{2.10b}$$

The angle Ψ_0 is called the "aperture (half-)angle" of the paraboloid, i.e. the angle between the axis and the "edge-ray" from F_P. The "depth" Δ of the paraboloid is given by

$$\Delta = \frac{{d_P}^2}{16f} = f\left(\frac{d_P}{4f}\right)^2 = f\left(\tan\frac{\Psi_0}{2}\right)^2. \tag{2.11}$$

The following relations also hold

$$\sin\psi = \frac{r/f}{1+(r/2f)^2}, \tag{2.12}$$

and

$$\cos\psi = \frac{(r/2f)^2-1}{(r/2f)^2+1}. \tag{2.13}$$

The formulas for the secondary reflectors in spherical coordinates are

hyperboloid (Cassegrain) $\quad r_s = (c^2 - a^2)\,/\,(a + c\cos\psi)$ (2.14)

ellipsoid (Gregorian) $\quad r_s = (c^2 - a^2)\,/\,(c - a\cos\psi)$ (2.15)

where c and a are the usual parameters describing the conic sections (see above).

The Cassegrain system is characterised by the **magnification factor m**, connected to the eccentricity $e = c/a$ of the hyperboloidal secondary by the relations

$$m = (e+1)\,/\,(e-1) \tag{2.16a}$$

or, equivalently,

$$e = (m+1)\,/\,(m-1) \tag{2.16b}$$

The "equivalent paraboloid" of the Cassegrain antenna is given by

$$\tan\frac{\phi}{2} = \frac{r}{2mf}$$

from which follows

$$\tan\frac{\Phi_0}{2} = \frac{d_p}{4\,m\,f}\,, \qquad (2.17)$$

where Φ_0 is the opening half-angle of the secondary reflector seen from the Cassegrain focus. Hence we have

$$m = \tan\left(\frac{\Psi_0}{2}\right) / \tan\left(\frac{\Phi_0}{2}\right) \qquad (2.18)$$

Further

$$c = \frac{d_s}{4}\,(\cot\Psi_0 + \cot\Phi_0), \qquad (2.19)$$

where c is the "focal length" and d_s the diameter of the hyperboloid. The distance between primary and secondary focus of the Cassegrain system is $f_c = 2\,c$ and the distance from primary focus to secondary vertex is $l = c - a$ (see Fig. 2.1). The following relations hold

$$f_c = 2\,c \qquad (2.20a)$$

$$l \equiv c - a = c\,(e-1)/e \qquad (2.20b)$$

$$a = c/e = c\,(m-1)/(m+1) \qquad (2.20c)$$

In the special case where the secondary focus coincides with the vertex of the primary reflector (i.e. $f_c = f$*), we have the situation that* Eq. *(2.17) can be written as* $\tan(\Phi_0/2) = d_s/4f$*, leading to the simple expression* $m = d_p/d_s$.

Table 2.1. Geometry of the ALMA antenna

Parameter	Symbol	Magnitude
Prime reflector diam.	d_p	12 m
Primary focal length	f_p	4.8 m
Primary focal ratio	f_p / d_p	0.4
Second reflector diam	d_s	0.75 m
Cassegrain focal ratio	f_s	8.0
Cass magnificaion	m	20
Distance between foci	f_c	6.177 m

Throughout this text, we shall illustrate the discussion with numerical examples. As the basic antenna for these we take the element antenna of the Atacama Large Millimeter Array (ALMA). The basic geometrical parameters of this Cassegrain antenna are assembled in Table 2.1. [Mat.2.1]

■ 2.2. Geometry of aberrations

For an optimal performance of the antenna, it is important that the system is properly focussed. For a prime focus reflector this means localizing the phase center of the feed in the focus of the primary paraboloid. In a Cassegrain system, the most important criterion is the coincidence of one of the foci of the hyperboloidal secondary reflector with the primary focus of the paraboloid. The location of the feed in the secondary focus is far less critical, as we shall show below. In this section we develop the geometrical formulae, which describe an axial or lateral defocus, that is a deviation of the feed or secondary reflector from the focal point along the axis of symmetry or perpendicularly to it. Later we shall use these to calculate the effects of defocus on the beam characteristics and the pointing of the antenna.

The general case of an arbitrary shift of the feed from the true focal position (the defocus δ) can be separated in two components: a shift along the reflector axis (axial defocus) δ_a and one perpendicular to the axis (lateral defocus) δ_l. We call ψ the angle between the reflector axis and a ray from the focus to a point at the surface at radius r. We now want to calculate the path-length difference between such a ray and the central, on axis, ray for both lateral and axial defocus. We can then incorporate the resulting phase error function in the basic radiation integral to calculate the radiation pattern and beam parameters of the defocused system. It is obvious that axial defocusing will cause a pathlength error which is independent of the azimuthal coordinate χ of the reflector aperture, while the pathlength due to lateral defocus in the azimuthal plane $\chi = \chi_0$ will be proportional to $\cos(\chi - \chi_0)$. In most of what follows we shall assume that the defocus is small with respect to the focal length of the reflector, i.e. $\delta / f << 1$, so that normally we can neglect terms of order $(\delta / f)^2$ and higher.

2.2.1. Lateral defocus

We treat first the case of a lateral defocus of the feed, or the secondary reflector, from the focal point. The situation is illustrated in Fig. 2.3; the defocus is denoted δ, omitting the subscript for lateral. Remember that the pathlength error will be dependent on the azimuthal aperture coordinate χ and we assume that the feed is moved in the plane where $\chi_0 = 0$. We shall derive below the full pathlength error, but later concentrate our calculations of the beam characteristics in the plane of defocus, where the effects are of course most pronounced. Applying the cosine rule to the triangle PFF', where P is a point on the reflector surface at radius r and the angle PFF' = $\pi/2 - \psi$, we have

$$\rho'^2 = \rho^2 + \delta^2 - 2\delta\rho \sin\psi \cos\chi,$$

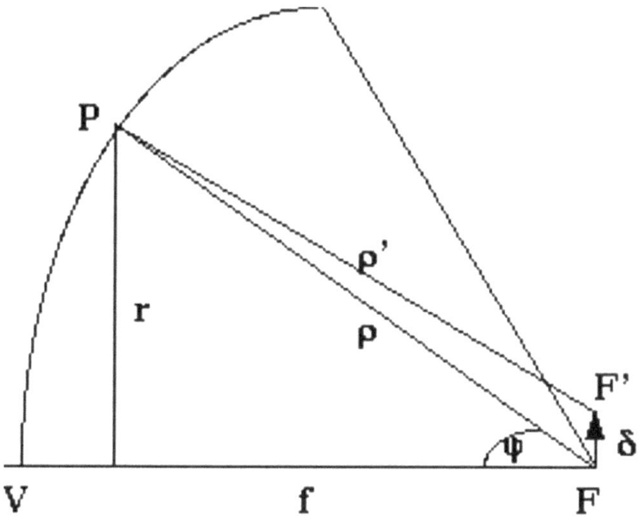

Fig. 2.3. The geometry of the lateral defocus in the plane of feed translation.

and using the series development for the square root $\sqrt{(1+x)} = 1 + \frac{x}{2} - \frac{x^2}{8} + ...$, we obtain

$$\rho' = \rho \left(1 - \frac{\delta}{\rho} \sin\psi \, \cos\chi + \frac{\delta^2}{2\,\rho^2} - \frac{\delta^2}{2\,\rho^2} \sin^2\psi \, \cos^2\chi\right),$$

where we ignore terms of order higher than $(\delta / \rho)^2$. Thus the pathlength difference Δ_l is

$$\Delta_l = \rho' - \rho = -\,\delta \, \sin\psi \, \cos\chi + \frac{\delta^2}{2\rho} - \frac{\delta^2}{2\rho} \sin^2\psi \, \cos^2\chi \, . \qquad (2.21)$$

Using several of the formulae for the description of the parabola in Section 2.1, we can eliminate both ρ and ψ from Eq. (2.21) to obtain the pathlength error Δ_L over the aperture as function of r, resulting from a lateral defocus δ in the plane χ=0,

$$\Delta_l = -\frac{\delta}{f}\, r\cos\chi + \frac{\delta}{4\,f^3}\, r^3 \cos\chi - \frac{\delta^2}{8\,f^3}\, r^2 - \frac{\delta^2}{2\,f^3}\, r^2 \cos^2\chi + \frac{\delta^2}{2\,f} . \qquad (2.22)$$

The terms in this equation represent some of the well known *aberrations* in optical instrument theory (see e.g. Born & Wolf, 1980). The first term is the *distortion*, which radio engineers call beam *tilt* or *squint*. It causes a shift of the beam maximum to an off-axis angle without disturbing the beam shape. The second term is the *coma* effect, also linear in δ but proportional to r^3. It causes a beam shift in the opposite direction by a smaller amount than the first term and moreover introduces an asymmetric beam distortion with a strong sidelobe on one side, the *coma lobe*. The third term is the *field curvature*, quadratic in r and independent of χ. It is reminiscent of an axial defocus, but the term's influence is not identical to that, because it is of second order in δ and inversely proportional to f^3 instead of f^2(see below). The fourth term is called *astigmatism*, characterised by features in the beam which are four-fold over the aperture. Astigmatism is sometimes an important aberration and we shall give it some attention later. Finally, the last term is independent of the integration variable and can be dropped. For a full description of these aberrations, together with pictorial illustrations, we refer to the standard work by Born and Wolf (1980, Ch. 5 and 9).

It is useful to return shortly to the field curvature term. It essentially reflects the fact that the surface in the focal region on which the image is most sharp (to borrow a term from optics) is not flat but curved. This surface is known as the Petzval surface, who derived the radius of this surface (see Born & Wolf, 1980, Ch. 5.5.3). Thus the optimum location of a laterally displaced feed will involve a correction to the axial position. Ruze (1965) gives the following corrective formula for the axial feed displacement δ_a needed to place a laterally displaced feed (by δ_l) on the Petzval surface:

$$\delta_a = \delta_l^{\,2} \,/\, 2\,f.$$

Thus the optimum locus of an off-axis feed lies on a paraboloid with a "focal length" half of that of the main reflector and the off-axis feed must be moved slightly away from the paraboloid's vertex.

Returning to the discussion of pathlength error, we note that the pathlength error of the central ray is $\rho' - \rho = \delta^2 \,/\, (\rho' + \rho) \approx \delta^2 \,/\, 2\,\rho$, which is equal to the second term in Eq. (2.21). Thus the phase error over the aperture, being proportional to the difference of the pathlength ρ'(r) and ρ'(0), is normally approximated by the first term of Eq. (2.21) only. From the geometry relations we obtain

$$\sin\psi = \frac{r}{\rho} = \frac{r}{2\,f}\,(1-\cos\psi) = \frac{r}{f}\left(\frac{1}{1+\tan^2(\psi/2)}\right) = \frac{r}{f}\left(\frac{1}{1+(r/2\,f)^2}\right).$$

As we assume that $(\delta\,/\,f) << 1$, the terms in δ^2 and higher can safely be neglected and we are left with a practical formula for the pathlength error due to lateral defocus of the form

$$\Delta_l = \delta\,\frac{r}{f}\left(\frac{1}{1+(r/2\,f)^2}\right)\,\cos\chi \qquad (2.23)$$

This function is illustrated in Fig. 2.4 [Mat.2.2] for the geometry of the earlier example (Table 2.1). We see that the error steadily increases from the center to reach a value almost as much as the lateral defocus at the edge of the reflector. It is easy to see that for a deep reflector with $f\ /\ d\ =\ 0.25$, where the focus lies in the aperture plane, the pathlength difference towards the edge is just equal to the defocus.

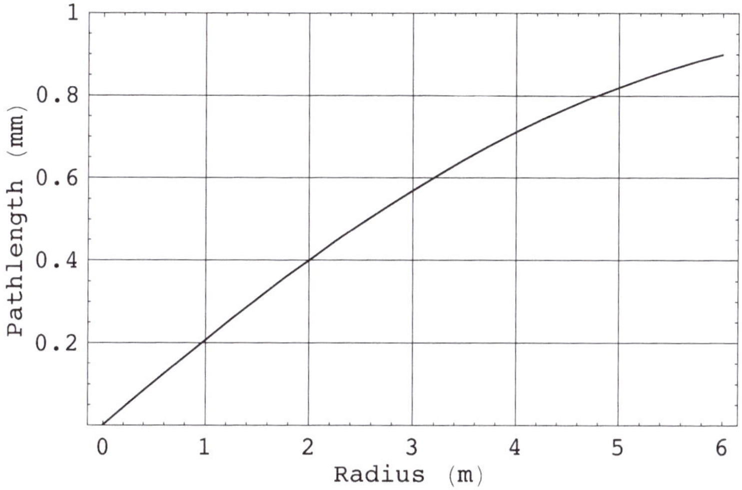

Fig. 2.4. Pathlength error in mm for 1 mm lateral defocus as function of aperture radius in m for a reflector with a focal ratio of 0.4.

To check the accuracy of the approximations in the above analysis, we can obtain exact expressions for ρ' and ρ by applying Pythagoras' Law to the pertinent triangles in Fig. 2.3 and calculating the difference in *Mathematica*, as shown in Fig. 2.5 [Mat.2.3] The difference is very small; for a defocus of 3 mm the error in Eq. (2.23) is less than 1 μm!

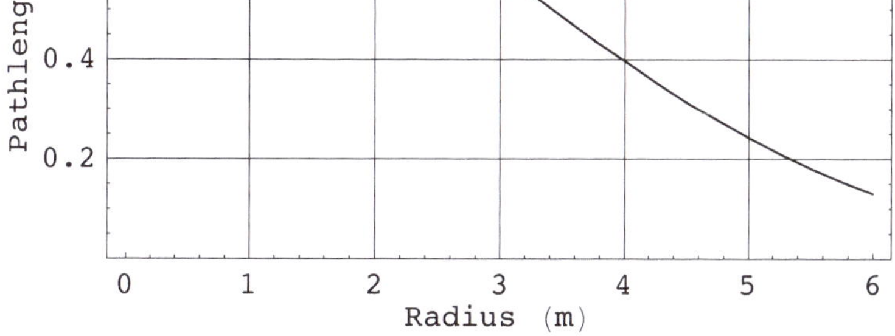

Fig. 2.5. Difference in the pathlength error between an exact calculation and the approximation of Eq. (2.23). For a defocus of 3 mm, the difference is less than one micrometer.

2.2.2. Axial defocus

We turn now to a discussion of axial defocus in a prime focus situation, depicted in Fig. 2.6. The defocus is denoted by δ (omitting the subscript for axial), assumed to be small with respect to the focal length, and we calculate the path-length difference from the defocused feed to the aperture plane between the on axis ray and the ray leaving the feed at an angle ψ with respect to the paraboloid's axis. Clearly the path length change of the central ray in the defocused case is δ. Using the geometrical relations for the parabola of Section 2.1, the path-length from F to V and thence to the aperture plane at a radius **r** is found to be

$$\rho_c = f + z(r) = f + \frac{r^2}{4f} = f\left(1 + \tan^2\left(\frac{\psi}{2}\right)\right).$$

From F to a point P on the reflector we have the relation

$$\rho = \frac{2f}{(1+\cos\psi)} = f\left(\frac{1+\cos\psi}{1+\cos\psi} + \frac{1-\cos\psi}{1+\cos\psi}\right) = f\left(1 + \tan^2\left(\frac{\psi}{2}\right)\right), \qquad (2.24)$$

which is identical to the value for the central ray, as the geometry of the parabola requires. Depending on the simplifications and approximations we allow in the derivation, we find slightly different representations for the path difference. We compare these now.

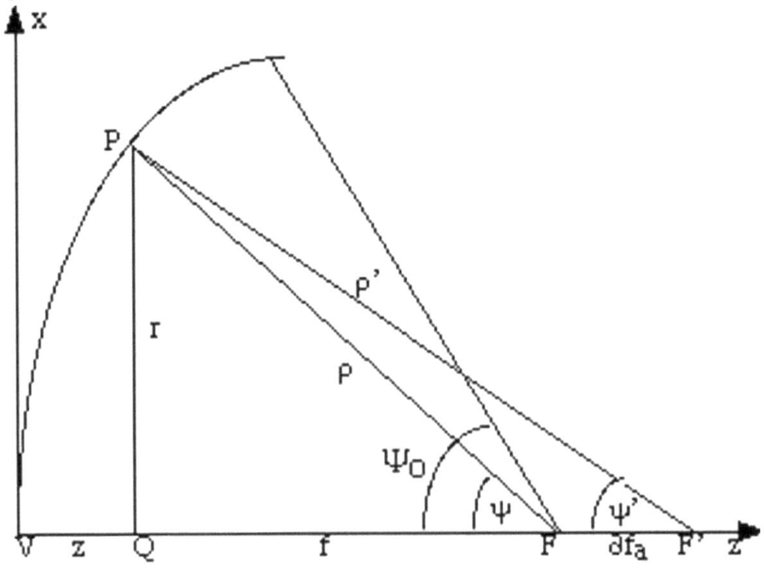

Fig. 2.6. Illustration of the geometry of the axial defocus

i). Assuming that $\cos\psi$ changes insignificantly with the focus shift, we find for the defocused ray $\rho' \approx 2(f+\delta)/(1+\cos\psi)$, leading to a path difference of

$$\rho' - \rho \approx 2\delta \,/\, (1 + \cos\psi).$$

Note that upon reflection both rays travel to the aperture of the antenna essentially parallel and we ignore the small length difference over that path. The phase error over the aperture is proportional to the difference in the pathlength change between any ray to a radius r and the central ray. Using several of the earlier geometric formulations for the paraboloid, we find for the pathlength difference Δ_a over the aperture, caused by an axial defocus δ

$$\Delta_a(r) = \delta\left(\frac{2}{1+\cos\psi} - 1\right) = \delta\,\frac{1-\cos\psi}{1+\cos\psi} = \delta\tan^2(\tfrac{\psi}{2}) = \delta\left(\frac{r}{2f}\right)^2 \qquad (2.25)$$

We see that, subject to the assumption that $\cos\psi$ changes insignificantly between the focused and defocused situation, the phase error due to axial defocusing is a simple quadratic function in the radial aperture coordinate r. The maximum error at the edge of the aperture with diameter **d** is thus

$$\Delta_a = \delta\left(\frac{d}{4f}\right)^2 = \delta\tan^2\left(\frac{\psi_0}{2}\right) \qquad (2.26)$$

ii). As an alternative, better approximation we can apply the cosine-rule to the triangle FPF'. We have

$$\rho'^2 = \rho^2 + \delta^2 - 2\rho\delta\cos(\pi-\psi) = \rho^2 + \delta^2 + 2\rho\delta\cos\psi,$$

from which we obtain the path length change of the ray to P

$$\rho' - \rho = \frac{2\rho\delta\cos\psi + \delta^2}{\rho' + \rho} \approx \delta\cos\psi + \frac{\delta^2}{2\rho} \approx \delta\cos\psi \qquad (2.27)$$

and hence the path error over the aperture is given by

$$\Delta_a(r) = \delta(\cos\psi - 1) = -\delta\,\frac{2\tan^2\frac{\psi}{2}}{1+\tan^2\frac{\psi}{2}} = -2\delta\,\frac{\left(\frac{r}{2f}\right)^2}{1+\left(\frac{r}{2f}\right)^2}. \qquad (2.28)$$

This result is obtained by ignoring the term in δ^2, assuming $\rho'+\rho = 2\,\rho$ and using Eq. (2.10a).

iii) If we apply Pythagoras' law to the triangles from F and F' to P, we obtain the following result

$$\Delta_a(r) = \sqrt{\left\{r^2 + \left(f - \frac{r^2}{4f} + \delta\right)^2\right\}} - \left\{f + \frac{r^2}{4f} + \delta\right\}. \tag{2.29}$$

Expansion of Eq. (2.29), calling the last term in curly brackets **a,** leads to:

$$\Delta_a(r) = \sqrt{a^2 - \frac{\delta\, r^2}{f}} - a \approx a\left(1 - \frac{\delta\, r^2}{2\,f\,a^2}\right) - a =$$

$$-\frac{\delta\, r^2}{2\,f\,a} = -\frac{\delta\, r^2}{2\,f}\;\frac{1}{f+\frac{r^2}{4\,f}+\delta} \approx -2\,\delta\,\frac{\left(\frac{r}{2f}\right)^2}{1+\left(\frac{r}{2f}\right)^2}.$$

In the last approximation we have retained the first two terms in the series expansion of the square root term and ignored the δ in the numerator of the second but last term. The result thus obtained is identical to Eq. (2.28).

The behaviour of Eqs. (2.25), (2.28) and (2.29) is illustrated in Fig. 2.7 [Mat.2.4], where we show the pathlength difference as function of aperture radius for an antenna with focal ratio 0.4 and a defocus of 1 mm.

The approximation of Eq. (2.28) is essentially indistinguishable from the exact result of Eq. (2.29) and hence is to be preferred over the approximation of Eq. (2.25). The difference between these solutions reaches only 1 μm for a defocus of 4 mm. Thus for all purposes Eq. (2.28) represents the path error adequately. If we adjust the maximum phase error at the aperture edge in Eq. (2.25) from $\tan^2(\frac{\psi_0}{2})$ to $(1 - \cos\psi_0)$, as given in Eq. (2.28), and just maintain the single quadratic dependence on r of Eq. (2.25), we find that this curve is reasonably close to the exact one and will often be acceptable for practical values of δ.

The discussion has dealt with the movement of the feed near the primary focus of the paraboloidal reflector. It is valid without any change for axial defocus of the feed in the secondary focus of a Cassegrain reflector configuration. We only have to consider the equivalent primary-fed paraboloid with a focal length of **m** times that of the real primary, where **m** is the magnification factor of the Cassegrain system, given by the expressions of Eq. (2.16) or (2.18). Also, it should be mentioned that an axial shift of the secondary reflector δ_s from the primary focus causes a phase error over the primary reflector as with a primary feed plus a small phase error over the secondary from its displacement with respect to the feed. We will discuss these aspects further in Section 4.3.3.

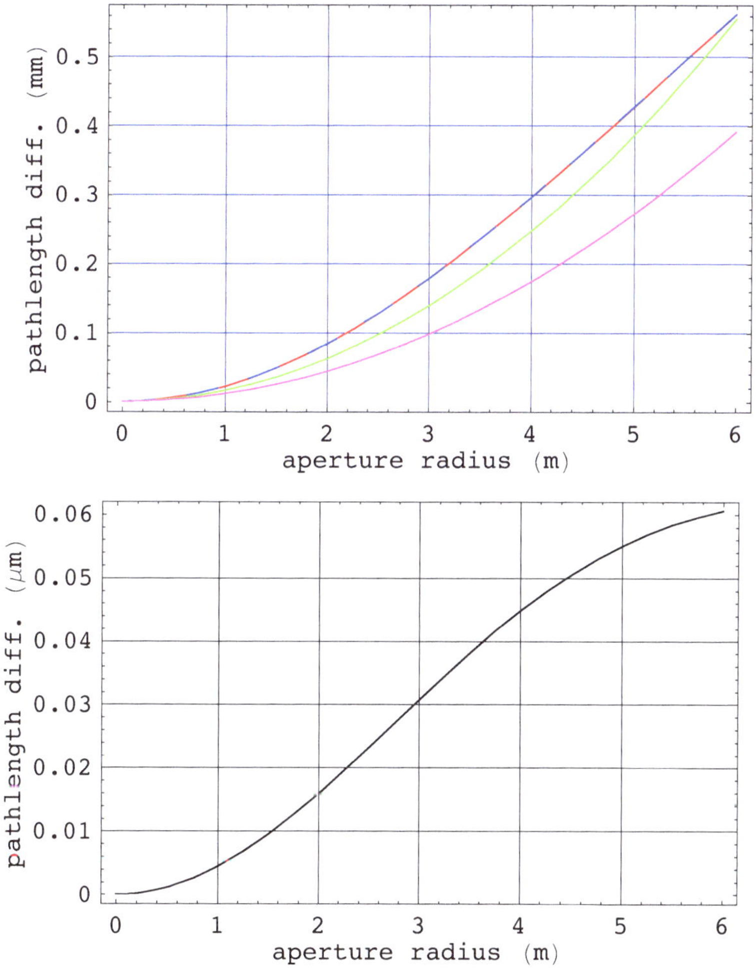

Fig. 2.7. Top: pathlength difference in mm from Eq. (2.29)-red, Eq. (2.28)-blue and Eq. (2.25)-magenta, as function of aperture radius for an axial defocus of 1 mm. The green curve is Eq. (2.25) but adjusted to the correct maximum path error at the aperture edge (see text). Note that the dashed blue curve effectively suppresses the exact red curve, showing their near identity. This is shown in the bottom plot of the difference between the red and blue curve with the vertical scale in micrometers.

This concludes the description of the geometry of the paraboloidal reflector antenna. In the following chapters we shall treat the electromagnetic characteristics of such antennas. This will include the calculation of the influence of defocus on the radiation pattern, for which the equations derived here will be needed. As we shall see, any defocus quickly deteriorates the beam pattern and being able to determine the correct focus is of great practical importance.

■ 2.3. The *Mathematica* Routines

This section contains the Mathematica routines used in the text of this chapter. They have been numbered Mat.2.x in the text and are identified by the same number in the first line of the expression.

```
Mat .2 .1 - ALMA antenna geometry;
dp = 12.0; f = 4.8; m = 20.; ds = 0.75;
Ψ0 = 2 ArcTan[dp / (4 f)] (180 / π)
Φ0 = 2 ArcTan[dp / (4 m f)] (180 / π)
e = (m + 1) / (m - 1)
m = (e + 1) / (e - 1)
fc = (ds / 2) (Cot[Ψ0 π / 180] + Cot[Φ0 π / 180])
l = fc (e - 1) / 2 e
```

```
Mat .2 .2 - lateral defocus pathlength error;
f = 4.8; χ = 0; δ = 0.001;
dl = (δ r / f) / (1 + (r / (2 f)) ^2);
Plot[1000 dl, {r, 0, 6}, PlotRange → {0, 1},
 Frame → True, GridLines → Automatic,
 FrameLabel → { "Radius (m)", "Pathlength (mm)"}]
```

```
Mat .2 .3 - Exact vs approx path error;
f = 4.8; δ = 0.003;
ρd = Sqrt[(r - δ) ^2 + (f - r^2 / (4 f)) ^2]; (defocus);
ρo = Sqrt[r^2 + (f - r^2 / (4 f)) ^2];        (in focus);
dl = (δ r / f) / (1 + (r / (2 f)) ^2);        (Eq .2 .23);
Plot[{10^6 (dl - (ρo - ρd))},
 {r, 0, 6}, PlotRange → {0, 1},
 Frame → True, GridLines → Automatic,
 FrameLabel → { "Radius (m)", "Pathlength (μm)"}]
```

```
Mat .2 .4 - Axial defocus, approximation comparison;
f = 4.8; δ = .001;
dl = -1000 (Sqrt[r^2 + (f - r^2/(4 f) + δ)^2] - (f + r^2/(4 f) + δ));
dm = 1000 δ 2 (r/(2 f))^2 / (1 + (r/(2 f))^2);
db = 1422 δ r^2/(4 f^2);
db1 = 1000 δ r^2/(4 f^2);
Plot[{ dl, db, dm, db1}, {r, 0, 6},
 Frame -> True, GridLines -> Automatic,
 PlotStyle -> {{RGBColor[1, 0, 0]}, {RGBColor[0, 1, 0]},
   {Dashing[{0.05, 0.05}], RGBColor[0, 0, 1]},
   {RGBColor[1, 0, 1]}}, FrameLabel ->
  {"aperture radius (m)", "pathlength diff. (mm)"}]
Plot[1000 { dm - dl}, {r, 0, 6}, Frame -> True,
 GridLines -> Automatic, FrameLabel ->
  {"aperture radius (m)", "pathlength diff. (μm)"}]
```

References

Born, M. and E. Wolf, *Principles of Optics*, 6th Ed., Oxford, Pergamon, 1980.

Ruze, J., Lateral feed displacement in a paraboloid, *IEEE Trans. Antennas Propagat.* **AP-13**, 660-665, 1965.

3. Electromagnetic theory of the reflector antenna

In this chapter we develop the general theory of the reflector antenna with a primary source of radiation (the "feed") in the focal point. We will not attempt to present a full rigorous theory of the radiation characteristics over the entire space around the antenna. Rather our primary interest is to find a representation of the radiation pattern in a limited solid angle around the boresight axis with a sufficient accuracy to reliably derive the important antenna parameters, as gain, sidelobe level, beam efficiency, etc. We include the influence of the finite distance of the field-point from the antenna. Thus we describe the radiation function both in the Fraunhofer (farfield) region at infinite distance, as well as in the Fresnel region, where the distance to the field point is finite.

We depart from Maxwell's equations and derive the wave equation which governs the fields radiated from a distribution of electric currents and charges in a volume of space bounded by a closed surface. We apply this to the case of a reflector antenna fed by a plane wave source in the focal point. The field from the feed induces surface currents on the reflector surface, which in turn are the sources of electro-magnetic radiation. The field strength in a point P of space, outside the antenna, is found from an integration of the currents on the illuminated reflector surface. Some approximations are introduced in this procedure. These lead to a representation of the radiated field as an integration of the field projected from the surface onto the aperture plane of the reflector; the so-called Kirchhoff-Helmholtz aperture integration. This approximation is sufficiently accurate for most applications in the area of large antennas, reckoned in terms of the wavelength used. The aperture integration is applied to a circular aperture and we obtain the radiation characteristics for field points both in the Fraunhofer and Fresnel regions of the antenna. The detailed discussion of these, including the effects of aberrations is reserved for the following chapter. We conclude this chapter with a short discussion of the interesting fact that the aperture field and the farfield are related by a Fourier Transformation. From a reciprocity relation it can also be derived that the spatial field distribution in the focal plane is identical to the field distribution in the farfield.

■ 3.1. Basic theory - Maxwell's equations

The description of any electromagnetic phenomenon must of course start with obeying the electromagnetic field equations of Maxwell. For a homogeneous, isotropic and linear medium, outside of the sources, these take the form

$$\nabla \times \mathbf{H} - \epsilon \frac{\partial \mathbf{E}}{\partial t} = \mathbf{j}, \qquad \nabla \cdot \mathbf{E} = \rho / \epsilon \tag{3.1}$$
$$\nabla \times \mathbf{E} + \mu \frac{\partial \mathbf{H}}{\partial t} = 0, \qquad \nabla \cdot \mathbf{H} = 0$$

Here **H** and **E** denote the magnetic and electric field strength, respectively, **j** is the electric current density, ρ the electric charge density, while ϵ and μ are the permittivity and permeability of the medium. Upon introduction of the vector potential **A**, defined by $\mathbf{H} = \frac{1}{\mu} \nabla \times \mathbf{A}$, one can easily show that **A** can be made subject to the following wave equation

$$\Delta \mathbf{A} - \frac{1}{c^2} \frac{\partial^2 \mathbf{A}}{\partial t^2} = -\mu \mathbf{j}. \tag{3.2}$$

As usual, we assume harmonic time dependence of the fields, written as exp(-i ω t), which reduces the time dependent members of Eq. (3.1) to

$$\nabla \times \mathbf{H} + i \omega \epsilon \mathbf{E} = \mathbf{j} \quad \text{and} \quad \nabla \times \mathbf{E} - i \omega \mu \mathbf{H} = 0. \tag{3.3}$$

Introducing the wave number k with $k^2 = \omega^2 \epsilon \mu$, the wave equation now becomes

$$\Delta \mathbf{A} + k^2 \mathbf{A} = -\mu \mathbf{j}\,. \tag{3.4}$$

A solution of Eq. (3.4) is

$$\mathbf{A} = \frac{\mu}{4\pi} \int_V \mathbf{j}\, \Psi \, dV, \tag{3.5}$$

where $\Psi = \exp(i k r) / r$ is the so-called Green function and the integration is carried out over the volume containing all current sources. Let us now divide space into a volume V_1, containing all sources and enclosed by a surface S, and an open, source-free volume in which the point of observation P, in which the field has to be determined, is located. It can be shown (e.g. Sommerfeld 1964, § 46) that the field in P can be found from an integration of the source-field values E_0 and H_0 over the volume V_1.

The magnetic field **H** in point P follows from the definition of **A** and Eq. (3.5)

$$\mathbf{H} = \frac{1}{4\pi} \nabla \times \int_{V_1} \mathbf{j}\,\Psi\, dV, \tag{3.6}$$

where the nabla operator applies to the coordinates of P. Substitution of the current density by an expression derived from Eq. (3.3) with source field quantities $\mathbf{E}_0$ and $\mathbf{H}_0$, and carrying out the vector operations on these fields, yields the following form for the magnetic field in point P (Schouten and de Hoop, 1952):

$$\mathbf{H} = \frac{1}{4\pi} \nabla \times \int_{V_1} [\nabla_0 \times (\Psi\, \mathbf{H}_0)]\, dV + \frac{1}{4\pi\, i\, \omega\, \mu} \nabla \times \nabla \times \int_{V_1} [\nabla_0 \times (\Psi\, \mathbf{E}_0)]\, dV. \tag{3.7}$$

The symbol ∇_0 operates on points within the source region V_1. By application of Gauss' theorem we can transform the expression to surface integrals:

$$\mathbf{H} = \frac{-1}{4\pi} \nabla \times \int_{S} [\mathbf{n} \times (\Psi\, \mathbf{H}_S)]\, dS - \frac{1}{4\pi\, i\, \omega\, \mu} \nabla \times \nabla \times \int_{S} [\mathbf{n} \times (\Psi\, \mathbf{E}_S)]\, dS,$$

where $\mathbf{n}$ is the unit normal to S pointing towards open space. Our point P will always be outside S and hence we can move the nabla operators under the integral sign. Because ∇ operates on the coordinates of P only, we can write

$$\nabla \times [\Psi\, (\mathbf{n} \times \mathbf{H}_S)] = -(\mathbf{n} \times \mathbf{H}_S) \times \nabla\, \Psi \quad \text{and}$$

$$\nabla \times \nabla \times [\Psi\, (\mathbf{n} \times \mathbf{E}_S)] = [(\mathbf{n} \times \mathbf{E}_S) \cdot \nabla]\, \nabla \Psi + k^2\, (\mathbf{n} \times \mathbf{E}_S)\, \Psi,$$

where we use Eq. (3.4) and the relation, valid for any vector **a,**

$$\nabla \times \nabla \times \mathbf{a} = \nabla\, (\nabla \cdot \mathbf{a}) - \Delta \boldsymbol{a}.$$

We assume that the medium outside S is a vacuum, where $k = \omega \sqrt{\mu_0\, \epsilon_0} = 2\pi/\lambda$. Thus we obtain

$$\mathbf{H}_P = \frac{1}{4\pi} \int_S \left\{ (\mathbf{n} \times \mathbf{H}_S) \times \nabla \Psi + i\, \omega\, \epsilon_0\, (\mathbf{n} \times \mathbf{E}_S)\, \Psi - \frac{1}{i\, \omega\, \mu_0} [(\mathbf{n} \times \mathbf{E}_S) \cdot \nabla]\, \nabla \Psi \right\}\, dS. \tag{3.8}$$

This result indicates that the field at any point P outside the region with sources bounded by the surface S can be found from a knowledge of the electric and magnetic fieldstrength over the surface S. Thus Eq. (3.8) can be considered as a representation of the electromagnetic "Huygens' Principle" (e.g. Baker and Copson, 1939). It can be shown (Fradin, 1961) that Eq. (3.8) yields correct results if the source fields vanish over part of S and the integration is performed over an open surface such as the aperture of a reflector antenna. Also, by a laborious derivation (see e.g. Fradin, 1961, p.66ff), the following equivalent expression to Eq. (3.8) can be found

$$\mathbf{H}_P \frac{1}{4\pi} \int_S \left(\mathbf{H}_S \frac{\partial \Psi}{\partial n} - \Psi \frac{\partial \mathbf{H}_S}{\partial n} \right) dS + \frac{1}{4\pi i \omega \mu} \int_L \nabla\Psi \; \mathbf{E}_S \cdot dl - \frac{1}{4\pi} \int_L \Psi \; [\mathbf{H}_S \times dl] \; , \tag{3.9}$$

where the path of the line integrals is the closed edge of the open surface S. For a closed surface S the line integrals vanish and the relation reduces to the well-known Kirchhoff - Helmholtz equation. We shall return to this equation below. First we turn to a calculation of the current density on the surface of the reflector resulting from a primary source.

■ 3.2. The primary source and surface current density

Several choices are possible for the primary source. We shall limit ourselves here to the most widely used feed for a reflector antenna, viz a pyramidal or conical waveguide horn. We approximate the field in the horn aperture by a plane wave with H-polarisation along the y-axis, say. Starting with Eq. (3.8) we can calculate the magnetic field strength near the surface of the reflector by integrating the field distribution of the feed over its aperture area. In all practical cases the reflector will be in the farfield region of the feed which allows for some simplification of Eq. (3.8). Without writing down the entire analysis we present here the result. The field in a point Q near the reflector surface due to a primary feed in the origin which transmits a plane wave of strength H_S can be written as

$$\mathbf{H}_Q(\rho, \psi, \chi) = \frac{-i}{2\lambda} \frac{\exp(i k \rho)}{\rho} F(\psi, \chi)(1 + \cos\psi)(\mathbf{i}_\psi \cos\chi + \mathbf{i}_\chi \sin\chi). \tag{3.10}$$

See Fig. 3.1 below for the meaning of the geometry. We call the function $F(\psi, \chi)$ the "feed pattern" or "illumination function" and it is given by

$$F(\psi, \chi) = \int_{S_f} \mathbf{H}_S(x, y) \exp\{-i k \sin\psi (x\cos\chi + y\sin\chi)\} \, dx \, dy, \tag{3.11}$$

where the integration extends over the aperture of the feed in the x,y-plane. The field $\boldsymbol{H}_Q$ is transversely polarised (independent of ρ). Our next step is to calculate the current density on the surface of the reflector as a result of this incoming field. In general the current in a point on the reflector will be influenced not only by the incoming field at that point but also by interaction with the currents in other parts of the surface. When the reflector is "locally plane", i.e. the radius of curvature is much larger than the wavelength, the situation can be simplified significantly. For a perfectly conducting reflector surface the boundary conditions prescribe the surface current as

$$\mathbf{j} = \mathbf{n} \times \mathbf{H},$$

where $\mathbf{H}$ is the total magnetic field at the surface and $\mathbf{n}$ is the local unit normal pointing away from the reflector. Because we also have $\mathbf{H} = \mathbf{H}_i + \mathbf{H}_r$ and $\mathbf{n} \times \mathbf{H}_i = \mathbf{n} \times \mathbf{H}_r$, where the subscripts indicate incoming and reflected field, respectively, we arrive at

$$\mathbf{j} = 2\,(\mathbf{n} \times \mathbf{H}_i). \tag{3.12}$$

By substitution of Eq. (3.10) into Eq. (3.12) we obtain the full expression for the surface current. With the aid of the relations of Eqs. (2.2, 2.3 and 2.13) we can derive the following expression:

$$\mathbf{j}_Q = -2\,\mathbf{H}_Q \left\{ \mathbf{i}_y \cos\tfrac{\psi}{2} + \mathbf{i}_z \sin\tfrac{\psi}{2} \sin\chi \right\}, \tag{3.13}$$

where the magnitude of the magnetic field H_Q follows from Eq. (3.10) as

$$\mathrm{H}_Q = -\frac{i}{2\lambda}\,\frac{\exp(ik\rho)}{\rho}\,F(\psi,\chi)\,(1+\cos\psi)\ . \tag{3.14}$$

■ 3.3. Surface current integration

Now that we have an expression for the current density at the illuminated side of the reflector surface, we can proceed with the calculation of the scattered field by the reflector. The current on the shadow side of the reflector is assumed to vanish and the assumed infinite conductivity of the reflector assures the existence of surface currents only. Departing from Eq. (3.6) and using Eq. (3.3) we see that the volume integral can be transformed into a surface integral to yield

$$\mathbf{E}_P(R,\theta,\phi) = \frac{1}{4\pi\omega\epsilon}\,\nabla\times\Big[\nabla\times\int_S \mathbf{j}_Q\,\frac{\exp(ikr)}{r}\,dS\Big], \tag{3.15}$$

where the field point P has coordinates (R, θ, ϕ) (see Fig. 3.1) and the integration is performed over the illuminated side of the reflector only. The nabla operator works on the field point P. Substitution of Eqs. (3.13-14) leads to

$$\mathbf{E}_P = \frac{-1}{8\pi^2}\sqrt{\frac{\mu}{\epsilon}}\int_S \nabla\times\nabla\times\left\{\frac{\exp i\,k(r+\rho)}{r\,\rho}\left(\mathbf{i}_y\cos\frac{\psi}{2} + \mathbf{i}_z\sin\frac{\psi}{2}\sin\chi\right)\right\} F(\psi,\ \chi)\,(1+\cos\psi)\,dS. \qquad (3.16)$$

We shall not write down the tedious, but straightforward, procedure of the repeated rotation operation on the expression in curly brackets. The general expression in curvilinear coordinates can be found in Stratton (1941, p.50). After transformation of the unit vectors $\mathbf{i}_y$ and $\mathbf{i}_z$ into the spherical coordinate system (R, θ, ϕ) by application of Eqs. (2.1 and 2.3) we obtain

$$\mathbf{E}_P = \frac{-1}{8\pi^2}\sqrt{\frac{\mu}{\epsilon}}$$
$$\int_S F(\psi,\ \chi)\,(1+\cos\psi)\,\frac{\exp i\,k(r+\rho)}{r\,\rho}\Big[\cos\frac{\psi}{2}\Big\{-2\,\mathbf{i}_R\sin\theta\,\sin\phi\left(\frac{i\,k}{r}-\frac{1}{r^2}\right) +$$
$$\mathbf{i}_\theta\cos\theta\,\sin\phi\left(k^2+\frac{i\,k}{r}-\frac{1}{r^2}\right)+\mathbf{i}_\phi\cos\phi\left(k^2+\frac{i\,k}{r}-\frac{1}{r^2}\right)\Big\} +$$
$$\sin\frac{\psi}{2}\sin\chi\Big\{-2\,\mathbf{i}_R\cos\theta\left(\frac{i\,k}{r}-\frac{1}{r^2}\right)-\mathbf{i}_\theta\sin\theta\left(k^2+\frac{i\,k}{r}-\frac{1}{r^2}\right)\Big\}\Big]$$
$$dS\ . \qquad (3.17)$$

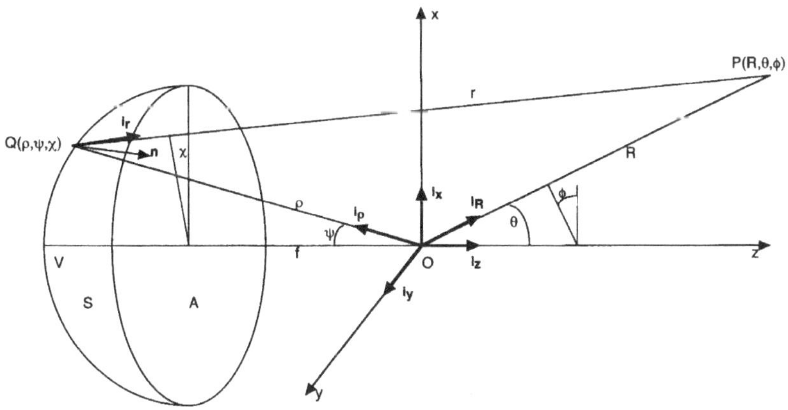

Fig.3.1. Geometry and definition of coordinate systems for the paraboloidal reflector.

The first term in square brackets results from the y-component (transverse component) of the surface current density on the reflector, while the second component is due to the "longitudinal" z-component of the current. This equation can be considerably simplified in essentially all practical cases, where we have r >> λ and reflector diameter d >> λ. In that case we can ignore the terms $\left(\frac{ik}{r} - \frac{1}{r^2}\right)$ compared to k^2 and, with $k = 2\pi/\lambda$, Eq. (3.17) reduces to

$$\mathbf{E}_P = \frac{-1}{2\lambda^2}\sqrt{\frac{\mu}{\epsilon}} \int_S F(\psi,\ \chi)(1+\cos\psi)$$
$$\frac{\exp ik(r+\rho)}{r\rho}\left[\cos\tfrac{\psi}{2}\,(\mathbf{i}_\theta \cos\theta \sin\phi + \mathbf{i}_\phi \cos\phi) - \mathbf{i}_\theta \sin\tfrac{\psi}{2} \sin\theta \sin\chi\right] dS. \tag{3.18}$$

As in Eq. (3.17), the first term in square brackets is produced by the y-component of the surface current, the second by the z-component. The field due to the y-component is independent of the azimuthal surface coordinate χ. For a point on the axis of the paraboloid ($\theta = 0$), the z-component vanishes and the field has the polarisation vector $\mathbf{i}_y$. This can be seen using Eqs. (2.1 and 2.3). The field does not have a component in the **r**-direction; hence it is transverse. For the integration over the surface we need to select two of the three coordinates (ρ, ψ, χ) and eliminate the third from Eq. (3.18). Normally it is most convenient to integrate over the full 2π azimuthal angle χ and the polar angle ψ over the range determined by the geometry of the reflector. Thus we eliminate ρ from Eq. (3.18). From the discussion in Chapter 2 and referring to Fig. 3.1, we can derive

$$\cos\psi' = \mathbf{i}_\rho . \mathbf{i}_R = \cos\psi \sin\theta \cos(\chi - \phi) + \sin\psi \cos\theta$$

and applying the cosine rule to the triangle OQP in Fig.3.1 we find

$$\begin{aligned} r^2 &= R^2 + \rho^2 - 2\rho R \cos\psi' \\ &= R^2 + \rho^2 - 2\rho R \{\cos\psi \sin\theta \cos(\chi - \phi) + \sin\psi \cos\theta\} \end{aligned} \tag{3.19}$$

In all practical circumstances we will have R >> ρ. Thus for the amplitude term of Eq. (3.18) we can derive from Eq. (3.19)

$$\frac{1}{r} = \frac{1}{R}\left\{1 + \frac{\rho}{R}\cos\psi' - \frac{\rho^2}{2R^2} + \ldots\right\} \approx \frac{1}{R},$$

and from the paraboloid geometry (Eq. (2.4)) we have

$$\frac{1}{\rho} = \frac{1+\cos\psi}{2f}.$$

In the exponential term of Eq. (3.18), the phase term, we apply Newton's iteration formula for the square root of a number to Eq. (3.19). If $x^2 = N$ and we choose an estimate x_i for $\sqrt{N}$, then $x_{i+1} = \frac{1}{2}\left(x_i + \frac{N}{x_i}\right)$ is correct to twice the number of significant figures compared to x_i. If we choose $r_1 = R$ as a first estimate, we find as the next better approximation

$$r \approx R - \rho \cos\psi' + \frac{\rho^2}{2R}. \tag{3.20}$$

This form, up to the second order in ρ, is used in the description of the Fresnel field of the antenna. We shall deal with this in more detail later. In the usual farfield region, the quadratic term can be ignored.

At this point we introduce a new variable (see Fig.3.1 and Eq. (2.10))

$$a = \frac{4f}{d} \tan\frac{\psi}{2} = \frac{2b}{d} \quad (0 \leqslant a \leqslant 1),$$

which is the normalised radius of the projection of point Q on the surface onto the aperture plane. With this variable we obtain for the surface element in Eq. (3.18)

$$dS = \frac{d^2}{4} \sec\frac{\psi}{2} \; a\, da\, d\chi. \tag{3.21}$$

For the phase term we obtain from the above approximations

$$r + \rho = R + \rho(1 - \cos\psi') = R + 2f\,\frac{1-\cos\psi'}{1+\cos\psi}$$

$$= R + 2f\left\{\cos^2\frac{\theta}{2} - \sin\theta\cos(\chi-\phi)\tan\frac{\psi}{2} + \sin^2\frac{\theta}{2}\tan^2\frac{\psi}{2}\right\}.$$

With the elimination of the variable ψ by introducing variable a this expression becomes

$$r + \rho = R - \frac{da}{2}\sin\theta\cos(\chi - \phi) + 2f\cos^2\frac{\theta}{2} + \frac{1}{2f}\left(\frac{da}{2}\right)^2 \sin^2\frac{\theta}{2}. \tag{3.22}$$

Introducing these approximations in Eq. (3.18) we obtain the following integral

$$\mathbf{E}_P = \frac{-1}{4\lambda^2}\,\frac{d^2}{f}\,\sqrt{\frac{\mu}{\epsilon}}\,\frac{\exp(i\,k\,R)}{R}$$
$$\int_0^1\!\int_0^{2\pi} F(a,\ \chi)\exp\Big[i\,k\Big\{-\frac{d\,a}{2}\sin\theta\,\cos(\chi-\phi)+\frac{(d\,a)^2}{8\,f}\sin^2\frac{\theta}{2}\Big\}\Big]$$
$$\Big[\cos\frac{\psi}{2}\,(\mathbf{i}_\theta\cos\theta\sin\phi+\mathbf{i}_\phi\cos\phi)-\mathbf{i}_\theta\sin\frac{\psi}{2}\sin\theta\,\sin\chi\Big]\,a\,da\,d\chi\,, \quad (3.23)$$

where the illumination function is written as $F(a, \chi) = F(\psi, \chi)\cos^3\frac{\psi}{2}$. This results from the elimination of the terms in ψ from the variables ρ and dS. Also we have ignored the phase term $2\,f\cos^2\frac{\theta}{2}$, which is independent of the integration variables and normally small compared to R.

■ 3.4. Aperture integration, Kirchhoff-Helmholtz diffraction

We now simplify the situation further as follows. Suppose we have a reflector which concentrates the reflected rays from the primary source in a half space, as for instance is the case with a paraboloid of revolution in the geometrical optics approximation. The closed surface on which the currents must be known can now be chosen as an infinite plane, close to the illuminated side of the reflector, and a hemispherical cap of infinite radius enclosing the reflector with all its surface currents. The point of observation lies outside this region. On the infinite plane we describe a closed finite curve which circumscribes the set of "reflected rays", i.e. we assume geometric optics reflection of the primary field at the reflector. This reflected field is projected upon the infinite plane and the area within the closed curve circumscribing the rays is called the ***aperture*** of the reflector, defined as the area within the projection of the shadow line on the reflector along the reflected rays onto the infinite plane. The field outside the aperture on the plane is taken to be zero and the contribution of the infinite cap vanishes by virtue of the radiation condition. Thus the integral over a closed surface reduces to that over a finite plane area. This is known as the *aperture field integration method.* As a last approximation we assume that the aperture is large in terms of the wavelength and that our interest in the diffraction pattern is restricted to a relatively small angular region around the maximum. From Eq. (3.13) with Eq. (3.14) we recall that the surface current has no component in the x-direction. If we project the surface current distribution onto a plane perpendicular to the z-axis (the "aperture plane", just defined), the contribution from the z-component vanishes and we have a linearly polarised field. We can deduce (Silver, 1949, p. 418-19) that the projected electric field in the aperture $\boldsymbol{E}_A$ is related to the current by

$$\mathbf{E}_A = -\,0.5\sqrt{\frac{\mu}{\epsilon}}\,\sec\Big(\frac{\psi}{2}\Big)\,\mathbf{j}_Q(y).$$

Applying Eq. (3.13) to this relation and inserting also Eq. (3.14) we find for the aperture field

$$\mathbf{E}_A = \sqrt{\frac{\mu}{\epsilon}}\, H_Q\, \mathbf{i}_y = \frac{-i}{2\lambda}\sqrt{\frac{\mu}{\epsilon}}\,\frac{\exp i k \rho}{\rho}\, F(\psi,\ \chi)\,(1 + \cos\psi)\,\mathbf{i}_y \tag{3.24}$$

Thus we see that the aperture field distribution is analogous to the angular distribution of the source ("feed") pattern as denoted in Eq. (3.14). Because of the paraboloid's geometry, the term $\exp(i k \rho)/\rho$ does not influence the phase function over the aperture compared to the feed pattern. The amplitude distribution over the aperture is however somewhat weaker toward the edge compared to the feed pattern, because the spherical wave from the feed travels a longer path to the reflector edge than to its center. This effect is known as the "free-space taper" of the illumination function. We shall return to this in more detail in the next chapter.

We continue now with the detailed treatment of the aperture integration method. Let us choose as the aperture plane the (x, y)-plane through the focal point O (Fig. 3.1 and 3.2). The aperture A is thus the projection of the reflector rim onto this plane. The electric field in this plane is given by Eq. (3.24). We now have ψ =0 and hence $\cos\psi' = \sin\theta\cos(\chi-\phi)$. Also the distance from the field point P to a point in the aperture follows from Eq. (3.20) as

$$r = R - b\sin\theta\cos(\phi - \chi) + \frac{b^2}{2R}, \tag{3.25}$$

where b is the radial coordinate in the aperture, thus b = d a / 2.

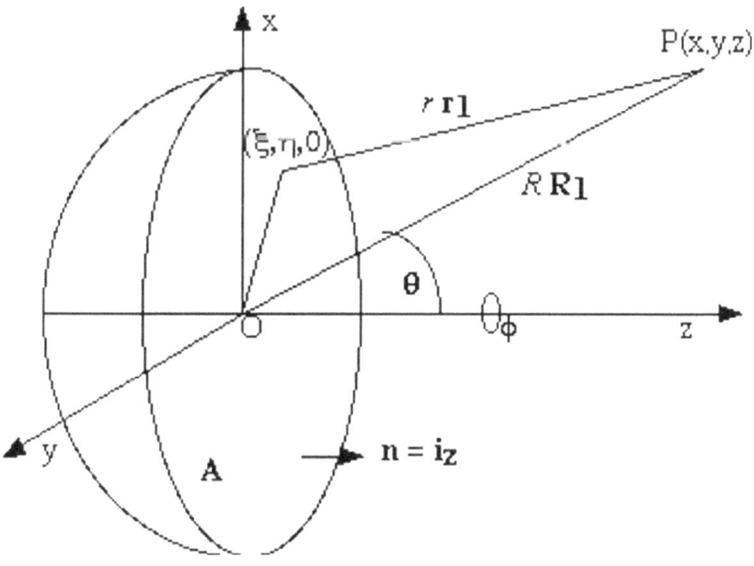

Fig. 3.2. Geometry of the aperture integration method; P is the field point.

We recall that the method assumes that the aperture and the distance to the field point are both large with respect to the wavelength and that the calculation is restricted to relatively small angles θ about the beam axis. In practice reliable results are obtained up to angles incorporating several sidelobes. For sufficiently large values of the distance R the second term in the exponent can be ignored. In that case we are dealing with *Fraunhofer diffraction* and we refer to the resulting field as the farfield pattern of the antenna. In cases where this term cannot be neglected, we are in the region of *Fresnel diffraction* and the point P is said to be in the nearfield region of the antenna. In the following sections we examine these two cases in more detail. We maintain the second order term here and arrive at the simplified form of Eq. (3.23)

$$\mathbf{E}_P = \frac{d^2}{4}\,\frac{\exp(i\,k\,R)}{R}\,(\mathbf{i}_\theta \cos\theta \sin\phi + \mathbf{i}_\phi \cos\phi)$$
$$\int_0^1 \int_0^{2\pi} F(a,\ \chi)\exp\left[i\,k\left\{-\frac{d\,a}{2}\sin\theta\,\cos(\chi-\phi)+\frac{(d\,a)^2}{8R}\right\}\right]a\;d\,a\,d\,\chi. \tag{3.26}$$

Here the z-component has vanished and we have neglected the small term of $\sin^2\frac{\theta}{2}$ in the curly brackets of Eq. (3.23). As a last step we notice that for small values of θ, i.e. close to the z-axis (the beam axis), the vector expression in front of the integral sign approaches the unit vector $\mathbf{i}_y$, equal to the polarisation of the aperture field. Thus we have reduced the treatment to the so-called *scalar aperture integration* and the radiation integral is written

$$\mathbf{E}_P(\theta,\ \phi) = \frac{d^2}{4}\,\frac{\exp(i\,k\,R)}{R}$$
$$\int_0^1 \int_0^{2\pi} F(a,\ \chi)\exp\left[i\,k\left\{-\frac{d\,a}{2}\sin\theta\,\cos(\chi-\phi)+\frac{(d\,a)^2}{8R}\right\}\right]a\;d\,a\,d\,\chi\,. \tag{3..27}$$

■ 3.5. The farfield approximation (Fraunhofer region)

In this and the following section we shall examine the behaviour of the radiation integral of Eq. (3.27) in more detail. In the **farfield** situation, the field point P is so far away (in principle at infinity) that the vectors $\boldsymbol{R}_1$ and $\boldsymbol{r}_1$ are essentially parallel. For the distance r we can limit ourselves to the linear part of Eq. (3.25)

$$r = R - b\sin\theta\ \cos(\phi-\chi), \tag{3.28}$$

where b = d a / 2 is the radial coordinate and χ the angular coordinate in the aperture plane. The radiation integral Eq. (3.27) is simplified to

$E_P(\theta, \phi) =$

$$\frac{d^2}{4} \frac{\exp(ikR)}{R} \int_0^1 \int_0^{2\pi} F(a, \chi) \exp\left[-ik\frac{d\,a}{2} \sin\theta \cos(\chi - \phi)\right] a\,d\,a\,d\,\chi. \tag{3.29}$$

We write the illumination function as $F(a, \chi) = A(a, \chi) \exp[\Phi(a, \chi)]$, where $A(a, \chi)$ is the amplitude function and $\Phi(a, \chi)$ the phase function. Normally, the illumination function will be rotationally symmetric and in a perfect reflector geometry the phase function will be a constant. In that case the integration over χ can be performed to yield the Bessel function of the first kind and order zero (see, e.g. Jahnke - Emde, 1945, p. 149). Thus the following expression is obtained:

$$f(\theta, \phi) = \frac{\pi d^2}{2} \int_0^1 A(a)\, J_0\left(\frac{k\,d}{2} a \sin\theta\right) a\,d\,a\,. \tag{3.30}$$

Here we have dropped the term in R, which is irrelevant for the radiation pattern function, which we write as $f(\theta, \phi)$ for *field-pattern*. We also introduce the new variable $u = \frac{k\,d}{2} \sin\theta$ for the beam angle. For a uniform illumination ($A(a) = 1$), Eq. (3.30) can readily be integrated to yield

$$f(\theta, \phi) = \frac{\pi d^2}{4}\, 2\, \frac{J_1(u)}{u} = \mathcal{A}\, \Lambda_1(u). \tag{3.31}$$

where J_1 is the Bessel function of the first kind and order one, and Λ_1 is the so-called Lambda function of first order (see Jahnke-Emde, 1945, p. 180). The Lambda function is defined as

$$\Lambda_n(u) = n!\, \frac{J_n(u)}{(u/2)^n}\,,$$

with $J_n(u)$ the Bessel function of the first kind and order n. $\mathcal{A}$ represents the area of the aperture. Because we are mainly interested in the functional form of $f(\theta)$, we omit the constant $\mathcal{A}$ from now on. For the assumed uniform illumination function the radiation function is circularly symmetric in the variable ϕ.

For illumination functions of the form $A(a) = (1 - a^2)^p$, $p = 0, 1, 2, \ldots$ the resulting farfield function can be written in terms of Lambda functions as

$$f(u) = \frac{\Lambda_{p+1}(u)}{p+1}\,. \tag{3.32}$$

An often used illumination function, which represents a typical feed pattern well, is the "quadratic on a pedestal"-function $A(a) = 1 - (1 - \tau)\, a^2$, where $0 \le \tau \le 1$.

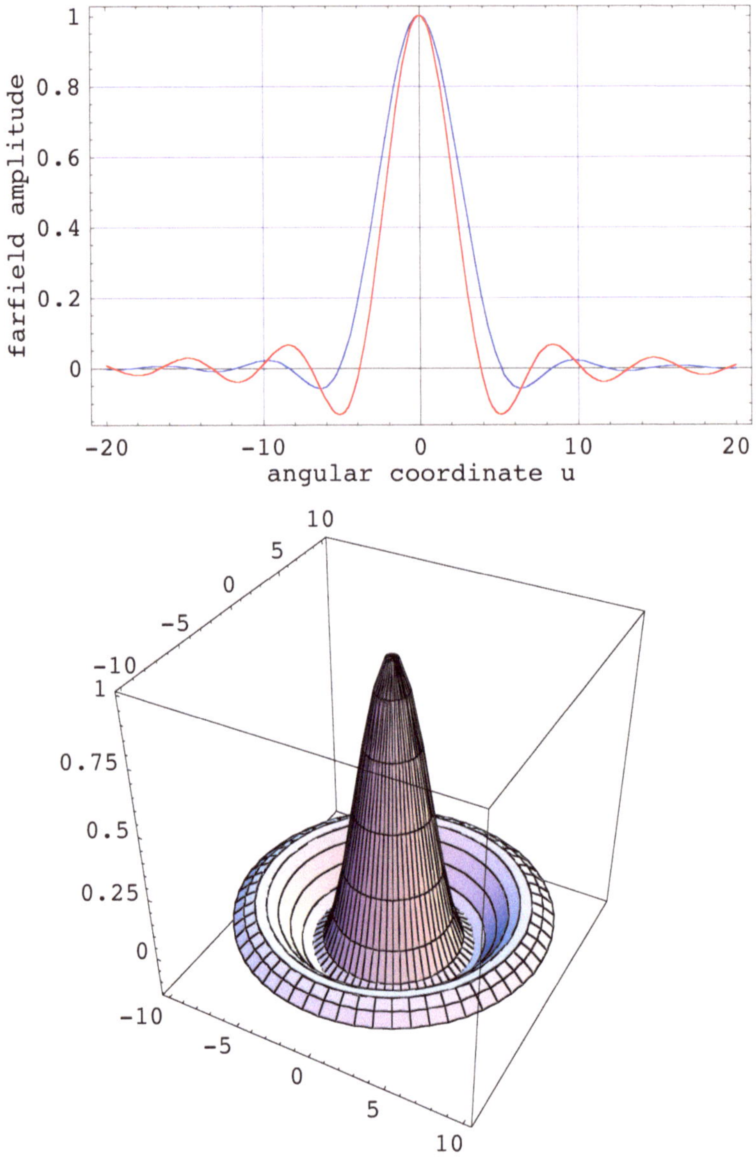

Fig. 3.3. The upper figure shows the one dimensional pattern for the uniform (red) and fully tapered (blue) illumination function. The lower figure is a three dimensional rendering of the uniformly illuminated case. It is the lambda function as given in Eq. (3.31).

The quantity τ determines the level of illumination at the edge of the aperture (the pedestal, also called the "taper"). We shall treat this in more detail in Chapter 4.

We now illustrate these results with some plots obtained with *Mathematica*. First we show the calculation for uniform (τ = 1) and quadratic (τ = 0) illumination by integrating Eq. (3.30). The routine [Mat.3.1] outputs the resulting functional form, where the appearance of the Lambda functions is obvious. The plots of Fig.3.3 show that the uniform illumination (red) leads to a narrower beam with higher sidelobes than the quadratic illumination (blue). Clearly by choosing the height of the pedestal, we can influence the shape of the radiation beam. This will be discussed in more detail in the next chapter. Figure 3.3 shows amplitude patterns with negative sidelobes. The square of the functions produces the usual *power pattern*, also known as the Airy pattern, consisting of a bright central core (the main beam or Airy disk) surrounded by dark and bright rings; these are the so-called sidelobes of the diffraction pattern (Fig.3.4), [Mat.3.2] The normalised power pattern of the uniformly illuminated aperture, also called the *gain function*, thus has the form

$$g(u) \equiv f^2(u) = \left\{\frac{2J_1(u)}{u}\right\}^2 \tag{3.33}$$

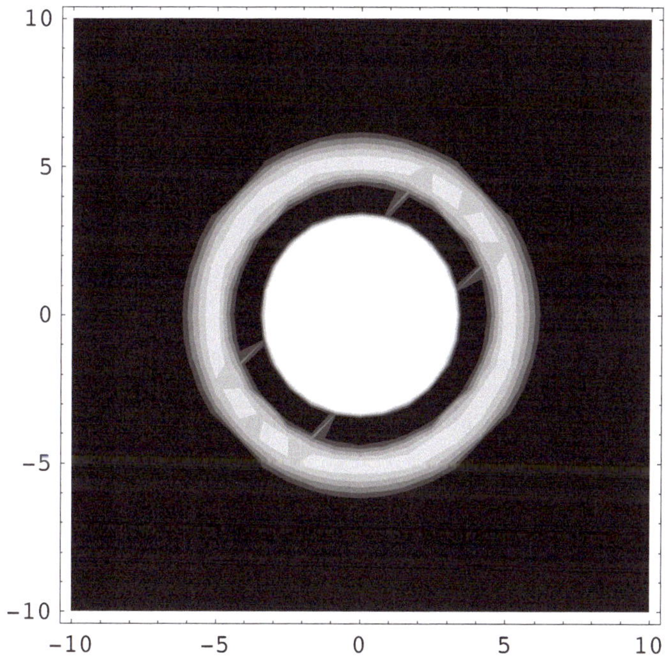

Fig. 3.4. Contour plot of the radiation beam of a uniformly illuminated aperture in the Fraunhofer region, also known as the Airy disc. The little spikes are due to the mathematical interpolations in the rendering of the plot. Two sidelobe rings are visible.

and is illustrated in the form of a contour plot in Fig. 3.4. The location of the minima (nulls) in the pattern is given by the roots of $J_1(u) = 0$ and that of the maxima by

$$\frac{d\,[J_1(u)/u]}{d\,u} = -\frac{J_2(u)}{u} = 0,$$

i.e. by the roots of $J_2(u) = 0$ (see e.g. Jahnke-Emde, p. 145, 1945). It is of interest for the later discussion of the concept *beam efficiency* to calculate the integrated energy over the pattern as function of the angular coordinate. Denoting $P(u_0)$ the fraction of the total power contained within the radius u_0, we have

$$P(u_0) = \int_0^{u_0} \int_0^{2\pi} g(u)\, u\, d u\, d\phi \propto \int_0^{u_0} \frac{J_1^{\,2}(u)}{u}\, d u. \qquad (3.34)$$

Using some of the recurrence relations for Bessel functions (see e.g. Jahnke-Emde, p.145) we obtain

$$\frac{J_1^{\,2}(u)}{u} = J_0(u)\,.\,J_1(u) - J_1(u).\,\frac{d\,J_1(u)}{d\,u}$$

$$= -J_0(u)\,\frac{d\,J_0(u)}{d\,u} - J_1(u)\,\frac{d\,J_1(u)}{d\,u} = \frac{-1}{2}\,\frac{d}{d\,u}\,\{J_0^{\,2}(u) + J_1^{\,2}(u)\}$$

which, considering that $J_0(0) = 1$ and $J_1(0) = 0$, leads to

$$P(u_0) = 1 - J_0^{\,2}(u_0) - J_1^{\,2}(u_0). \qquad (3.35)$$

Table 3.1. Parameters of the uniformly illuminated aperture

angular coord.	Intensity	Int.(dB)	Integrated energy
0	1.0	0	0.0
3.832	0.0	–	0.84
5.136	0.0175	–17.6	–
7.016	0.0	–	0.91
8.417	0.0042	–23.8	–
10.174	0.0	–	0.94

This expression was first derived by Rayleigh (1881) and is shown in Fig. 3.5 [Mat.3.3] One clearly sees the plateaux at the positions of the dark rings of the Airy pattern. If u_0 coincides with a null of the pattern we have $J_1(u_0) = 0$ and the power outside that radius simply is equal to $J_0^{\,2}(u_0)$. The main beam, up to the first null, contains 84 percent of the total radiated energy. Numerical data are assembled in Table 3.1.

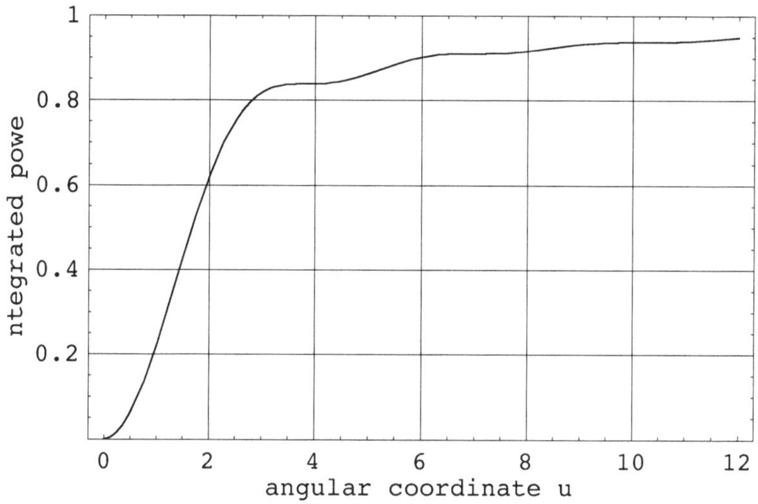

Fig. 3.5. The integrated power in the diffraction pattern as function of the angular coordinate. The plateaux near the minima in the pattern at u= 3.8 and 7.0 are clearly visible. The main beam contains 84 percent of the power.

■ 3.6. The nearfield approximation (Fresnel region)

In the **nearfield** region, which corresponds to the Fresnel region in optical diffraction (see e.g. Born and Wolf, 1980) we need to evaluate Eq. (3.25) without simplification. The radiation integral is thus given by Eq. (3.27), which we rewrite here without the irrelevant terms in front of the integral sign. As before, we consider a circular aperture.

$$f(\theta, \phi) = \int_0^1 \int_0^{2\pi} F(a, \chi) \exp\left[i k \left\{-\frac{d\,a}{2} \sin\theta \cos(\chi - \phi) + \frac{(d\,a)^2}{8R}\right\}\right] a \, da \, d\chi. \quad (3.36)$$

For a rotationally symmetric aperture distribution $A(a)$ with a constant phase term, the integration over χ results in

$$f(\theta) = 2\pi \int_0^1 A(a) J_0 \left(k \frac{d\,a}{2} \sin\theta\right) \exp\left(i k \frac{(d\,a)^2}{8R}\right) a \, da. \quad (3.37)$$

These integrals have been studied by Lommel (1884, 86) in his treatment of Fresnel diffraction at a circular aperture and the solution can be written in terms of Lommel functions of two variables (for details cf., Baars (1970) and Born & Wolf, Ch. 8.8 (1980)). Although Lommel's treatment is in itself interesting and moreover feasible in practice because of extensive tabulations of his functions (e.g. Dekanosidze,

1960), we prefer here to use *Mathematica* for the evaluation of Eq. (3.37), [Mat.3.4]. In the integration we must separate the exponential in its cosine and sine parts. Thus we obtain:

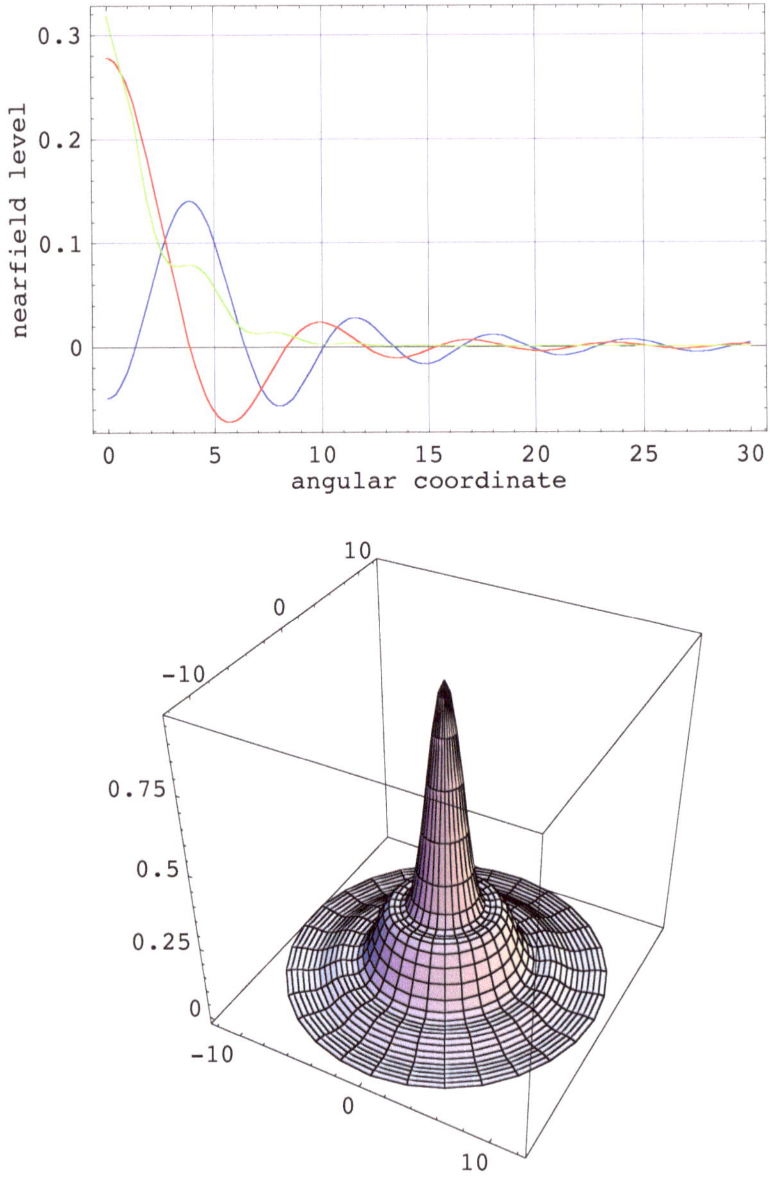

Fig.3.6. The red and blue curves are the sine and cosine-component, respectively, while the green curve is the "power pattern" (multiplied by a factor 4 for purpose of illustration). The distance from the aperture is 300 m. A 3-D picture of the power pattern is show in the lower part, where the "shoulder" near u = 3 is clearly visible.

$$f_c(\theta) = \int_0^1 J_0(u\,a)\,\cos\left[k\,\frac{(d\,a)^2}{8\,R}\right]\,a\,da\,,$$

and

$$f_s(\theta) = \int_0^1 J_0(u\,a)\,\sin\left[k\,\frac{(d\,a)^2}{8\,R}\right]\,a\,da\,.$$

In Figures 3.6 and 3.7 we show illustrations of the nearfield by numerically integrating Eq. (3.37). As before, we choose a normalised aperture radius, introduce the variable $u = k\,\frac{d}{2}\sin\theta$ and ignore the factor $2\,\pi$. We choose here a wavelength of

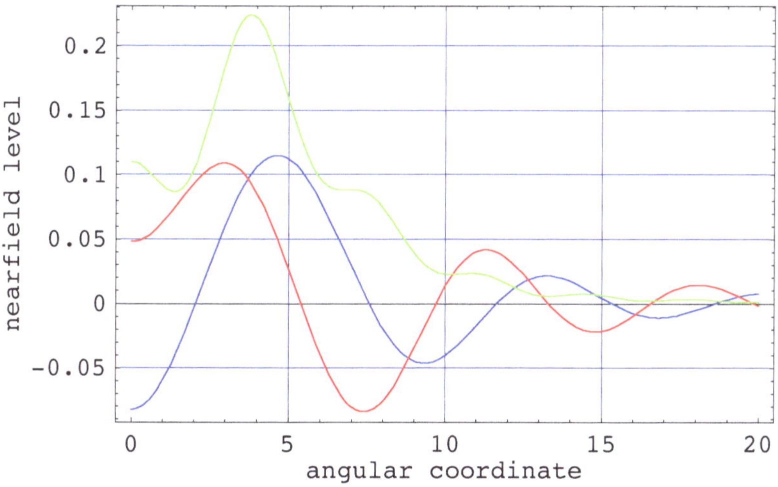

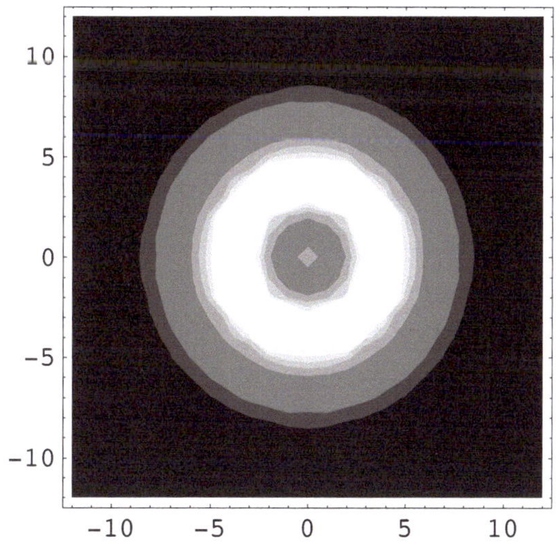

Fig.3.7. A nearfield calculation as in Fig.3.6 for a distance to the aperture of 200 m. The on-axis value is now suppressed by more than a factor 2 with respect to the maximum. The contour plot nicely indicates the bright ring near u = 3 - 4.

3 mm and evaluate the field function at a distance of R = 300 m and 200 m from the aperture, respectively. A value of u=30 corresponds with an angle of about 1 degree off boresight. The power pattern is the sum of the squares of the functions f_c and f_s. In the following calculation of figure (3.7) all parameters are the same but for the distance which has been reduced to 200 m [adjust Mat.3.4] We now notice that the intensity on the axis is much lower than at angle u = 4, indicating that the phase structure in the nearfield can lead to destructive interference on-axis.

It is interesting to look at the behaviour of the Fresnel region field level along the beam axis. We find that it oscillates strongly with minimum values of zero. The Bessel function in Eq. (3.37) drops from the integration, because for θ=0 we have $J_0(0) = 1$. For uniform illumination the effect is most strongly visible and Eq. (3.37) simplifies to:

$$f(0, R) = \int_0^1 \exp\left(i\, k \frac{a^2}{2R}\right) a\, da.$$

After separating the integral into its cosine and sine parts, each can be easily integrated. The power function, being the sum of the squares of the cosine and sine parts, has the form (ignoring irrelevant constants)

$$g(0, R) = 1 - \cos\left(\frac{k}{2R}\right), \tag{3.38}$$

which is illustrated in Fig.3.8, [Mat.3.5]. For a non-uniform aperture function the

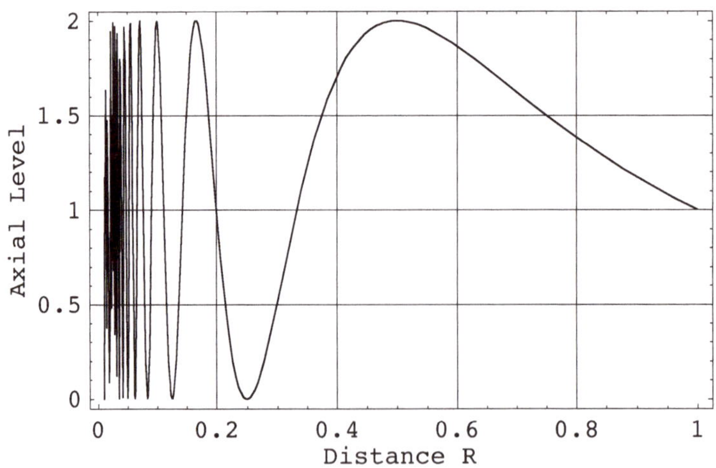

Fig.3.8. The power density on the beam axis as function of the distance from the aperture. At short distances the fast variation in phase causes a quickly oscillating function, which monotonically decreases only beyond the "Rayleigh distance" from the aperture.

destructive interference from the different parts of the aperture will not be complete and the axial power density will not oscillate to zero. To improve the visibility of the plot we have multiplied the axial value by R^{-2}. The peak axial power decreases with the distance squared, which is to be expected.

In the Fresnel region, the power is radiated more or less in a cylinder of a diameter about equal to that of the reflector. At the distance $d^2/2\lambda$, called the *Rayleigh distance*, the beam starts to expand into the farfield diffraction beam with an opening angle equal to the farfield beam width. We will return to this in more detail during the description of the dual-beam observing technique in Chapter 6.

■ 3.7 The Fourier Transformation relationship

Consider again Fig.3.2 with attention to the cartesian coordinates (x, y, z) of the field point P. We introduce the **direction cosines** of the field point P (u, v = sin θ cos ϕ, sin θ sin ϕ); see Ch. 2, Eqs. 2.1 and 2.3. Using these it is easy to show that the radiation integral of Eq. (3.29), written in cartesian coordinates, can be expressed as

$$f_P(u, v) = \frac{e^{ikR}}{R} \int\int F(\xi, \eta) \exp\{ik(\xi u + \eta v)\}\, d\xi\, d\eta . \tag{3.39}$$

We see that there exist a Fourier Transformation relationship between the field strength in point P, $f(u, v)$, and the field function in the aperture $F(\xi, \eta)$.

Ignoring the term in front of the integral sign, the **inverse Fourier transformation** can now be written as

$$F(\xi, \eta) = \int\int f(u, v) \exp\{-ik(u\xi + v\eta)\}\, du\, dv , \tag{3.40}$$

where the integration of the complex field $f(u, v)$ in principle has to be performed over a closed surface, surrounding the aperture. Thus a knowledge of the entire complex farfield pattern, **both in amplitude and in phase,** provides a description of the complex field distribution $F(\xi, \eta)$ over the aperture of the antenna, also in amplitude and phase.

In his standard book on antenna theory, Silver (1949) devotes an extensive discussion to this relationship (Ch. 6.3), and concludes that any practical application of this relationship is limited by the fact that the farfield pattern is only measurable in power. Thus the phase function of $f(u, v)$ will be arbitrary and hence the aperture distribution cannot be uniquely determined. It was Jennison (1966) who mentioned, in the appendix of his pocket book "Radio Astronomy", the same relation and its possible practical usefulness, pointing out that the amplitude and phase can both be measured with an interferometer. When Silver wrote his text in the mid forties, radio interferometry had not yet been developed. We will return to this important result when we discuss the holographic measurement technique for determining the geometrical shape of a reflector in Chapter 6.

As an example we present below the Fourier Transformation of a square aperture with tapered amplitude and constant phase distribution (Fig.3.9); the *Mathematica* expression is given in [Mat.3.6].

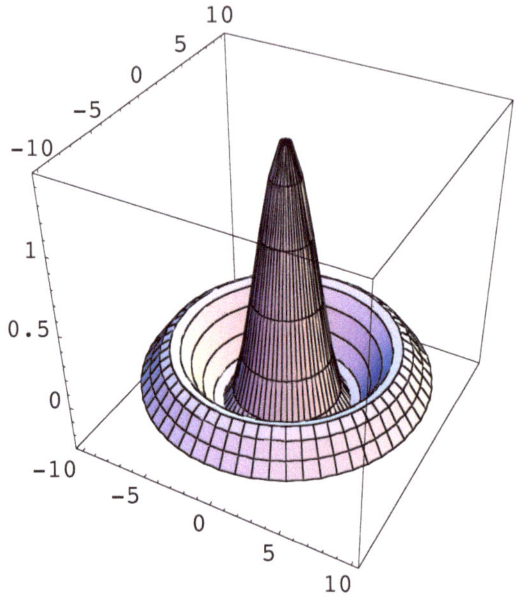

Fig.3.9. Three dimensional representation of the Fourier transformation of a rectangular, tapered aperture distribution.

The circular symmetric case of the circular aperture distribution discussed above is a special case of Fourier transformation. Such transformations are known as Hankel Transformation. Because of the azimuthal symmetry it reduces to a one-dimensional transformation. This can be demonstrated easily by rewriting the transform equation in polar coordinates. In the aperture plane we introduce $r^2 = x^2 + y^2$, in the transform plane (antenna beam) $w^2 = u^2 + v^2$. With $x + i\,y = r\,\exp(i\,\theta)$ and $u + i\,v = w\,\exp(i\,\phi)$ Eq. (3.39) can be transformed into

$$f(w) = \frac{e^{i\,k\,R}}{R} \int_0^\infty F(r) \Big[\int_0^{2\pi} \exp\{i\,k\,w\,r\cos(\theta - \phi)\}\,d\theta\Big]\,r\,d\,r = \frac{e^{i\,k\,R}}{R}\,2\,\pi \int_0^\infty F(r)\,J_0(2\,\pi\,w\,r)\,r\,d\,r\,. \tag{3.41}$$

This relation is identical to Eq. (3.30) and represents the radiation function of a circular aperture with amplitude distribution $F(r)$ and uniform phase distribution. The result of this Hankel Transformation is the well known Lambda function $\Lambda_1(w)$.

■ 3.8. Relation between farfield and focal region field

Finally we want to mention the relationship between the farfield radiation function and the field distribution in the focal plane of the reflector. The farfield pattern has been obtained from "transmitting" a spherical wave from the focal point, letting it being "diffracted" by the parabolic reflector to form the outgoing wave function, which exhibits a spherical wave front with a Lambda-function amplitude distribution. We can now turn the situation around and consider a set of waves of uniform amplitude and constant phase to fall upon the reflector from a range of angular directions with respect to the axis.. The induced surface currents will radiate "Huygens wavelets" towards the focal point, where they will be vectorially added to form the focal plane field distribution. Based on the general principle of reciprocity (see for instance Silver, 1949 or de Hoop and de Jong, 1974) the functional form of the focal plane distribution will be the same as that of the farfield pattern function. Thus we will find a central bright spot in the focal plane surrounded by rings of decreasing amplitude separated by nulls. The intensity distribution will be given by the Lambda function. The term "Airy disc" originates from this fact, as it is seen in the focal plane of a telescope observing a bright star. We will not discuss focal fields in detail. It will however be clear that this approach is of great value in the design of feed horns for reflector antennas. More on this subject can be found in the IEEE reprint collection, edited by Love (1976).

In the following chapters we shall use these results when we discuss the practicalities of determining the characteristics of large reflector antennas like radio telescopes and satellite communication ground stations.

■ 3.9. The *Mathematica* Routines

```
Mat .3 .1 - "Lambda function" beam;
ff = Integrate[(2 / (1 + t))
    (2 r (1 - (1 - t) r^2)) BesselJ[0, u r], {r, 0, 1}]
Plot[Evaluate[Table[ff, {t, 0, 1}],
   {u, -20, 20}, PlotRange → All,
   Frame → True, GridLines -> Automatic,
   FrameLabel → {"angular coordinate u",
     "farfield amplitude"}, PlotStyle →
    {{RGBColor[0, 0, 1]}, {RGBColor[1, 0, 0]}}]]
<< Graphics`SurfaceOfRevolution`
SurfaceOfRevolution[2 BesselJ[1, u] / u,
  {u, -10, 10}, BoxRatios -> {1, 1, 1}]
```

Out[7]= $$\frac{4\,t\,u\,\mathrm{BesselJ}[1,u] - 8\,(-1+t)\,\mathrm{BesselJ}[2,u]}{(1+t)\,u^2}$$

```
Mat .3 .2 - Contour plot of Lambda func;
u = (x^2 + y^2)^0.5;
j1 = (2 BesselJ[1, u] / u)^2;
ContourPlot[Evaluate[j1, {x, -10, 10},
  {y, -10, 10}, ContourLines → False]]
```

```
Mat .3 .3 - Integrated beam power;
Plot[1 - BesselJ[0, u]^2 - BesselJ[1, u]^2,
 {u, 0, 12}, PlotRange → {0, 1}, Frame → True,
 GridLines → Automatic , FrameLabel →
  {"angular coordinate u", "integrated power"}]
```

```
Mat .3 .4 - Fresnel region patterns;
λ = 0.003; k = 2 π / λ; R = 300;
fc = NIntegrate[
   r BesselJ[0, u r] Cos[k r^2 / (2 R)], {r, 0, 1}];
fs = NIntegrate[r BesselJ[0, u r] Sin[k r^2 / (2 R)],
   {r, 0, 1}];
fn = {fc, fs, 4 (fc^2 + fs^2)};
Plot[Evaluate[fn, {u, 0, 30}, PlotRange -> All,
  Frame → True, GridLines → Automatic, FrameLabel →
   {"angular coordinate", "nearfield level"},
  PlotStyle -> {{RGBColor[0, 0, 1]},
    {RGBColor[1, 0, 0]}, {RGBColor[0, 1, 0]}}]]
SurfaceOfRevolution[12 (fc^2 + fs^2),
 {u, 0, 12}, BoxRatios → {1, 1, 1}]
ContourPlot[Evaluate[(fc^2 + fs^2), {x, -10, 10},
  {y, -10, 10}, ContourLines → False]]
```

```
Mat .3 .5 - Fresnel region axial power;
λ = 2; k = 2 π / λ;
Plot[ 1 - Cos[k / (2 R)], {R, .01, 1}, Frame -> True,
 GridLines -> Automatic, PlotRange → All,
 FrameLabel -> {"Distance R", "Axial Level"}]
```

```
Mat .3 .6 - Fourier - transform exam. ;
x = . ; u = . ;
FourierTransform [
  (Sign [1 - x] + Sign [1 + x]) (1 - .25 x ^ 2), x, u]
SurfaceOfRevolution [%, { u, -10, 10},
 BoxRatios → {1, 1, 1}]
```

Out[45]= $\dfrac{\sqrt{\frac{2}{\pi}}\,(-1.\,u\,\mathrm{Cos}[u] + (1. + 1.5\,u^2)\,\mathrm{Sin}[u])}{u^3}$

References

Baars, J.W.M., *Dual-beam parabolic antennae in radio astronomy*, Groningen, Wolters-Noordhof, 1970.

Baker, B.N. and E.T. Copson, *The Mathematical Theory of Huygens' Principle*, Oxford Univ. Press, 1939.

Born, M. and E. Wolf, *Principles of Optics*, 6th Ed.,Oxford, Pergamon, 1980.

Dekanosidze, E.N., *Tables of Lommel's functions of two variables*, Oxford, Pergamon, 1960.

Fradin, A.Z., *Microwave Antennas*, Oxford, Pergamon, 1961.

Hoop, A.T. de and G. de Jong, Power reciprocity in antenna theory, *Proc. IEEE* **121**, 1051-1056, 1974.

Jahnke, E. and F. Emde, *Tables of functions*, 4th Ed., New York, Dover, 1945.

Jennison, R.C., *Introduction to Radio Astronomy*, London, Newnes, 1966.

Lommel, E., Die Beugungerscheinungen einer kreisrunden Öffnung und eines kreisrunden Schirmchens, *Abh. II Classe Bayer. Akad. der Wiss.* **15**, 233, 1884.

Love A.W. (Ed.), *Electromagnetic Horn Antennas*, New York, IEEE Press, 1976.

Rayleigh, Lord, On Images formed without Reflection or Refraction, *Phil. Magazine* **11**, 214-218, 1881. (Scientific Papers, Vol. 1, 513-518, New York, Dover, 1964)

Schouten, J.A and A.T. de Hoop, Het beginsel van Huygens bij Buigingsproblemen, *Tijdschrift Nederlands Radiogenootschap* **17**, 45-62, 1952.

Silver, S., *Microwave Antenna Theory and Design*, MIT Rad. Lab Series **12**, New York, McGraw-Hill, 1949.

Sommerfeld, A., *Optik*, Leipzig, Akad. Verlaggesellschaft, 3rd Ed., 1964.

Stratton, J.A., *Electromagnetic Theory*, New York, McGraw-Hill, 1941.

4. Antenna characteristics in practical applications

■ 4.1. Introduction

After the development of the basic equations, we now turn to application of these to practical situations in antenna theory, like calculations of efficiency, beamwidth and sidelobe level as well as the influence of defocus and other errors on the antenna's characteristics. We shall develop formalisms and present results which are of use in the daily work of the antenna engineer and the radio astronomer.

We first introduce as basic quantity the *aperture efficiency*, defined as

$$\eta_A = A \;/\; A_g \,, \tag{4.1}$$

where A is the maximum absorption area and A_g the geometrical area of the antenna aperture. It indicates the efficiency with which the radiation from a point source, for instance the transmitter on a communication satellite, is collected. The *aperture efficiency* is a parameter defining the sensitivity of the antenna. It is determined by a number of phenomena and hence it can be seen as the product of a number of separate "efficiency components". Following, for instance, Kraus (1966, Ch. 6.25b), we can write the aperture efficiency as the product of a number of individual components:

$$\eta_A = \eta_i \, \eta_s \, \eta_r \, \eta_p \, \eta_e \, \eta_f \, \eta_b \,, \tag{4.2}$$

where η_i = illumination efficiency of the aperture by the feed function ("taper")

η_s = spillover efficiency of the feed (and subreflector, if present)

η_r = radiation efficiency of the reflector surface (ohmic loss)

η_p = polarisation efficiency of the feed-reflector combination

η_e = surface error efficiency ("Ruze loss"), also called scattering efficiency

η_f = focus error efficiency (both lateral and axial defocus)

η_b = blocking efficiency due to quadripod, subreflector, other obstruction.

It is important to note here that by this definition all components of the final aperture efficiency are related to the geometrical area of the reflector aperture. In other words, each of them reduces the aperture area by an amount proportional to the individual efficiency magnitude.

Before we deal with these in a quantitative way, we give a short description of each of these components.

- ***illumination (taper)***: the most important component of the aperture efficiency is another basic quantity, the *illumination efficiency*, which is determined by the radiation pattern of the feed. It describes the degree to which the outer areas of the aperture are less effectively exploited as a result of the "weaker" illumination of that area. This so-called illumination *tape*r is chosen to reduce the nearby sidelobes and the spillover, as we have already seen in Chapter 3. Thus there is a relation between the illumination and spillover efficiency. The detailed choice of the taper depends on the required sidelobe level, the ratio of receiver temperature to the expected spillover contribution (which will be different for Cassegrain and prime focus systems) and the acceptable beam broadening and loss of sensitivity. Often a maximization of the ratio aperture area (proportional to the gain G) to system temperature (the G/T_s-ratio in communication engineering) is the goal, where T_s is the overall system temperature, which contains a spillover contribution.

- ***spillover***: this is the percentage of power, emitted by the feed, which "spills over" the edge of the reflector and hence is not used to form the diffraction beam. The spillover efficiency is defined as the percentage of feed-radiated power which falls within the boundary of the reflector. The spillover is directly related to the edge taper of the illumination function. The received spillover power is dependent on whether the "spilled" radiation comes mainly from the warm ground (prime focus) or cool sky (Cassegrain). It is also dependent on the elevation angle. It can be an important factor in low-noise systems.

- ***ohmic loss***: with a metallic reflector this factor is close to one, but at very high frequencies (submm wavelengths) and/or with a paint layer on the reflector, the losses may become significant (several percent).

-***polarisation***: the loss due to cross-polarisation is mainly dependent on the feed design, although some of it can be caused by the curved reflectors. It is often the least known factor, because it is not easy to calculate and difficult to measure accurately. We shall not be concerned with it further.

- ***surface errors***: the small scale, randomly distributed deviations of the reflector from the prescribed (paraboloidal) shape cause randomly distributed phase errors over the aperture. Their effect has been analyzed by Ruze (1952, 1966) and resulted in the well known "Ruze formula" for the efficiency loss due to these errors. Strictly speaking, the Ruze analysis is valid only in case the correlation length of the surface errors is much larger than the wavelength and much smaller than the reflector diameter. In practice, however, this requirement appears to be rather flexible. We shall treat this important aspect in detail later in this chapter.

- ***focus***: axial or lateral displacements of the feed from the focus cause large scale, systematic phase errors over the aperture, which normally are amenable to calculation. One tries to minimize these errors by (regular) determination of the optimum focus from test observations. A full discussion of aberrations caused by defocus will be presented below.

- ***blocking***: the central subreflector and its support structure cause a partial shadowing of the aperture, which leads to a loss of efficiency. The blocking area consists normally of two parts: the "plane wave blocking", equal to the projection of the structure onto the aperture plane, and the "spherical wave blocking", which is the shadow cast by the spherical waves traveling from the outer region of the reflector (outside the support penetration point) to the focus. Only if the support is attached at the rim of the main reflector will the spherical blocking component vanish. This parameter will also treated in detail in this chapter.

Summarising, the aperture efficiency is first of all determined by the illumination efficiency. The other errors which are distributed over the aperture will have a reduced influence on the aperture efficiency to the extent that they are diminished by the smaller illumination intensity in the outer aperture area. Their influence on the overall efficiency is "weighted" by the illumination function at the location where their effect occurs. For a realistic evaluation of the overall aperture efficiency such a weighting is permissible. For instance, it is well known that the influence of the generally larger structural deformations towards the edge of the reflector is diminished by the smaller illumination level and hence gives a reduced contribution to the gain loss. Similarly, we can weigh the blocking areas with the illumination function, which will yield a smaller effective blocking area. It should be noted that, after having applied the weighting, where applicable, and having computed the effective loss in aperture, in order to correctly introduce this in the efficiency formula, it must be treated as a loss of area compared to the geometrical aperture area of the real reflector.

An important section of this chapter contains the effects of a displacement of the feed from the correct focal point, which we call *defocus*. We shall see that already small values of defocus have a significant influence on antenna parameters like gain and sidelobe level. Curves and formulae for these are presented. We also treat the influence of the *blocking* of the aperture by the secondary reflector, the central receiver box and their support legs. The last is often in the form of a symmetrical *quadripod*. Finally we discuss the important case of random error in the profile of the reflector, originally treated by Ruze (1952), in some detail. The chapter is concluded with a short discussion of large scale aberrations, like astigmatism.

■ 4.2. Illumination efficiency, beam width, sidelobe level

4.2.1. Illumination efficiency ("taper")

The illumination function of the aperture $F(r, \varphi)$ is equivalent, but not identical, to the radiation pattern of the feed in the focal point. This function determines the illumination efficiency of the antenna, which is the ratio of the gain of the antenna to that of a uniformly illuminated aperture. The illumination function is normally characterized by the "edge taper", i.e. the level of the illumination at the edge of the reflector compared to that in the center. However, different functional forms of the illumination function with the same edge level can result in different values for the illumination efficiency. The edge taper is normally expressed in decibel with respect to the central illumination level; thus with τ the relative field strength at the aperture edge, the taper T (in dB) is $T = 20\log(\tau)$ dB, $(0 < \tau < 1)$.

The level of the feed function at the opening angle of the reflector is further reduced by the *"free-space taper"*, determined by the path length difference between the edge and central rays. The free-space taper T_f (in dB) is dependent on the focal ratio f/d and is given by the expression ([Mat.4.1] and shown in Fig. 4.1):

$$T_f = 20\log\left(\left(\sec\frac{\Psi_0}{2}\right)^2\right) \equiv 20\log\left(1 + \left(\frac{d}{4f}\right)^2\right). \tag{4.3}$$

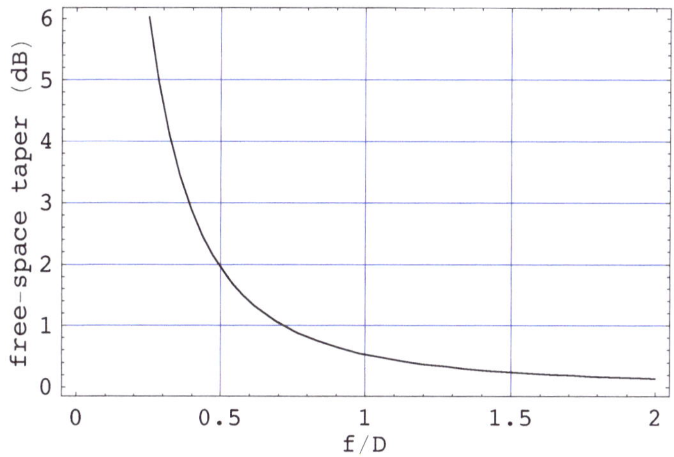

Fig. 4.1. The free-space taper in dB as function of the focal ratio of the reflector.

It is worthwhile noting that in the calculation of the free-space taper with Eq. (4.3) the focal ratio is that of the primary focus situation. Thus for a Cassegrain system with magnification m (see Ch. 2) the value of f in the above equation will be m times the focal length of the primary mirror. With typical values of m $\gtrsim$ 10, we see that normally the free-space taper for a Cassegrain is negligible.

The **illumination efficiency** η_i is defined (Silver, 1949) by the following equation:

$$\eta_i = \frac{\left\{\int F(r,\varphi)\, dA\right\}^2}{\int F^2(r,\varphi)\, dA}, \tag{4.4}$$

where the integration is extended over the aperture. We consider circularly symmetric amplitude illumination functions with constant phase front. Thus the integration over φ is trivial and the integration over r can be carried out in closed form for suitably chosen functions F(r). Here follow the results for two widely applied illumination functions, each with a freely chosen value of the level at the aperture edge, the edge taper T_e in dB. Note that this value includes the free-space taper.

i) the *gaussian* distribution with an edge taper T_e (in dB), is expressed by

$$F(r) = \exp\{-\alpha r^2\}, \tag{4.5}$$

where r is the normalised aperture radius and $\boldsymbol{\alpha = (T_e / 20) \ln 10}$. For T_e= -12 dB, we have $\alpha = 1.3816$.

Substituting Eq. (4.5) into Eq. (4.4) and performing the integration yields

$$\eta_i = \frac{2(1-e^{-\alpha})^2}{\alpha(1-e^{-2\alpha})}, \tag{4.6}$$

which for the selected taper of -12dB delivers an illumination efficiency $\boldsymbol{\eta_i = 0.866}$.

ii) the *quadratic on a pedestal* distribution, with an edge level τ and normalised to one in the aperture center, is given by

$$F(r) = \tau + (1-\tau)\{1-r^2\} \equiv 1-(1-\tau)\,r^2. \tag{4.7}$$

After substitution into Eq. (4.4) and carrying out the integration, we obtain

$$\eta_i = \frac{\{1-(1-\tau)/2\}^2}{\tau+(1-\tau)^2/3} = \frac{3(1+\tau)^2}{4(1+\tau+\tau^2)}, \tag{4.8}$$

where $\boldsymbol{\log(\tau) = T_e / 20}$, again T_e expressed in dB. This last relation [Mat.4.2] is shown in Fig. 4.2. For $T_e = -12$ dB we find $\boldsymbol{\eta_i = 0.893}$. This is near the value for the gaussian distribution, indicating that the illumination functions are rather similar, as illustrated Fig. 4.3 [Mat.4.3]. The quadratic function lies somewhat above the gaussian one and it results in a slightly higher illumination efficiency. The calculated spillover efficiencies are 0.95 and 0.75 for the quadratic and gaussian distribution, respectively. This strengthens the argument in favour of the quadratic function. In Fig. 4.4 we show the two functions of Eqs. (4.6) and (4.8) as function of the taper T_e in dB, indicating the higher efficiency of the quadratic distribution [Mat.4.4]. The quadratic distribution, going to zero at the reflector edge ($\tau = 0$), reduces the illumination efficiency to 0.75 times the maximum value for uniform illumination. For a

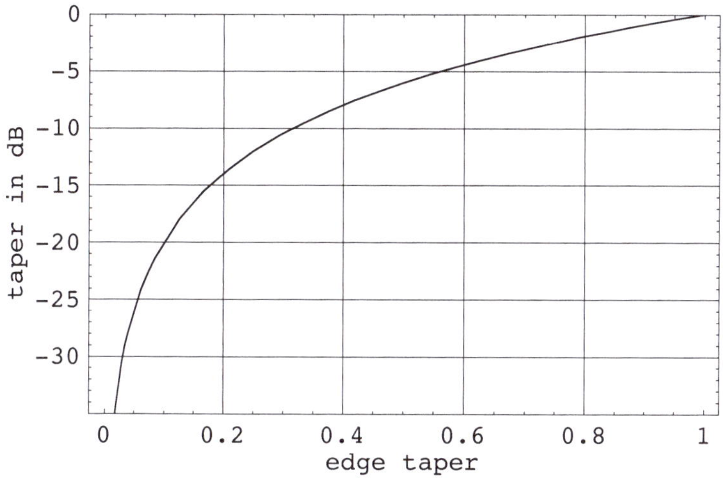

Fig. 4.2. The edge taper in dB against the value of τ.

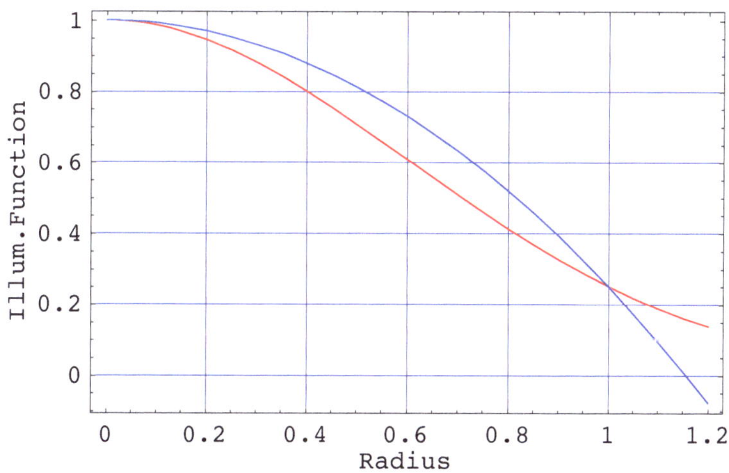

Fig. 4.3 The illumination function for the gaussian (red) and quadratic (blue) distribution and an edge taper of -12 dB. The quadratic illumination is somewhat more effective.

gaussian illumination with an edge value of about 2%, i.e. $\alpha \approx 4$, the illumination efficiency decreases to about 0.5. Obviously, to maximize the efficiency for a given edge taper, the quadratic distribution is preferable. Indeed, optimized feed horns exhibit a pattern, which is approaching the quadratic function, rather than the often assumed gaussian shape. For the commonly applied taper values between -10 and -20 dB, the difference is however limited to 10-15 percent.

Finally, note that these illumination functions implicitly contain the free-space taper. For instance, if the desired aperture edge taper is -15 dB and the reflector geometry introduces 3 dB of free-space taper, the feed must be designed to have a -12 dB amplitude decrease at the opening angle of the reflector. In the next section we discuss the influence of the illumination function on important antenna parameters as beamwidth and sidelobe level.

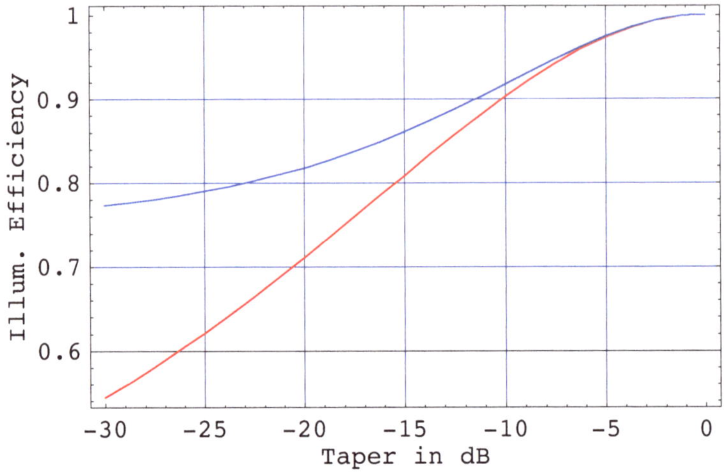

Fig. 4.4. The illumination efficiency as function of the edge taper (in dB) for the gaussian (red) and quadratic (blue) distribution.The quadratic illumination is more effective, especially at high taper values.

4.2.2. Beamwidth, sidelobe level and taper

In Chapter 3.5 we derived the normalised radiation function as (Eq. (3.29)

$$f(\theta,\ \phi) = \frac{d^2}{4} \int_0^1 \int_0^{2\pi} F(r,\ \chi)\, e^{i\,\Phi(r,\,\chi)}\ \exp\!\left[-i\,k\,\frac{d\,r}{2}\sin\theta\,\cos(\chi-\phi)\right] d\,r\ d\,\chi\,,$$

where $F(r,\ \chi)$ and $\Phi(r,\ \chi)$ are the amplitude and phase terms of the illumination function, respectively. When we assume a constant phase function over the aperture and introduce a rotationally symmetric amplitude illumination function (as e.g. Eq. (4.7)), the integration over χ can readily be performed and we obtain the rotationally symmetric radiation integral as

$$f(u) = \frac{\pi d^2}{2} \int_0^1 F(r)\ J_0(u\,r)\ r\ d\,r, \tag{4.9}$$

where $J_0(u\,r)$is the Bessel function of the first kind and order zero and the angular coordinate $u = (\pi\,d/\lambda)\sin\theta$, d being the diameter of the aperture. After introduction of the tapered aperture function according to Eq. (4,7), the integration over r yields (see Ch. 3.5)

$$f(u) = \left\{ \tau \, \frac{J_1(u)}{u} + 2(1-\tau) \, \frac{J_2(u)}{u^2} \right\}. \tag{4.10}$$

which can also be written in the form of Lambda functions as

$$f(u) = \left\{ \tfrac{\tau}{2} \, \Lambda_1(u) + \tfrac{1-\tau}{4} \, \Lambda_2(u) \right\}, \tag{4.11}$$

where the Lambda function is defined as $\Lambda_n(u) = n! \, \dfrac{J_n(u)}{(u/2)^n}$, with $J_n(u)$ the Bessel function of the first kind and order n. The Lambda functions are tabulated in Jahnke-Emde's "Tables of Functions", p.180ff, 1945. Thus, for uniform illumination (τ=1) we obtain the Lambda function of order one, while for full quadratic illumination (τ=0) the pattern is given by the Lambda function of order two. It is convenient to introduce a normalisation factor for the calculated radiation patterns as function of the taper parameter τ. This way, the computed pattern maximum will be one for any value of τ. Now, for $u = 0$ the two Lambda functions have the value one. Thus the normalisation factor n of Eq. (4.11) will obey the following expression:

$$n\left(\tfrac{\tau}{2} + \tfrac{1-\tau}{4}\right) = 1 \qquad \text{and hence} \qquad n = \frac{4}{1+\tau}.$$

The "**power pattern**" (also called **gain function**) ***g(u)*** of the antenna is the square of the function $f(u)$. Written in Bessel functions, the normalised power pattern thus becomes

$$g(u) = \left\{ \frac{4}{1+\tau} \left(\tau \, \frac{J_1(u)}{u} + 2(1-\tau) \, \frac{J_2(u)}{u^2} \right) \right\}^2. \tag{4.12}$$

The Bessel functions are directly available in *Mathematica*. The following figure (4.5) illustrates the function $g(u)$, expressed in logarithmic (dB) scale, for values τ = (0, step 0.2, 1) [Mat.4.5]. The narrowest curve pertains to the uniform illumination, the broadest to the quadratic illumination with edge level zero. This is qualitatively obvious, because the weaker illumination towards the aperture edge is equivalent to making the aperture smaller and hence the beam broader. From Fig. 4.5 we also see that the sidelobe level is smaller for the tapered aperture distribution. One can envisage this by considering that the superposition of the secondary "Huygens" wavelets is less effective because of the differing amplitudes, resulting in a smaller summation value.

The *Mathematica* routine also provides a table of the u-coordinates, where the functions are at the half-power level (- 3 dB), again in the order uniform illumination to fully tapered:

```
{{u → 1.61634}, {u → 1.64895}, {u → 1.69239},
 {u → 1.75311}, {u → 1.84397}, {u → 1.99442}}
```

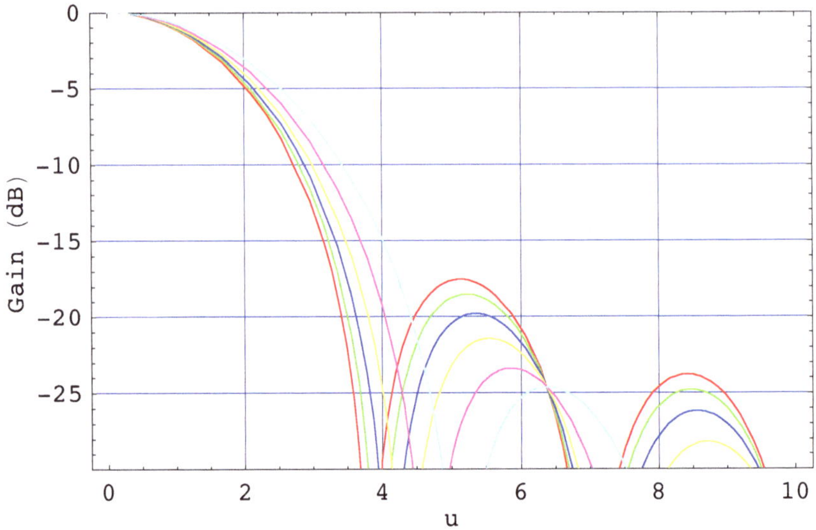

Fig. 4.5. Power pattern (in dB) of a circular aperture with quadratic illumination function and edge taper values of {1, step -0.2, 0}, from red, green, blue, yellow, magenta to cyan.

We can use these results to construct a curve of the half power beam width (HPBW) as function of the illumination taper τ. The half-power levels are found at the u-values listed above. We have

$$u_A = (\pi d \,/\, \lambda) \sin\theta_A \;\approx\; (\pi d \,/\, \lambda)\,\theta_A,$$

and

$$\theta_A \approx \frac{u_A\,\lambda}{\pi d}.$$

For the full HPBW we write $\Theta_A = \mathrm{b}\,(\lambda \,/\, \mathrm{d})$ and hence find for the factor $b = 2\,u_A \,/\, \pi$. Over the range $0.1 < \tau < 1$ the factor b can be approximated very well by the expression ([Mat.4.6], see also Fig. 4.6):

$$b \;=\; 1.269 \;-\; 0.566\,\tau + \;0.534\,\tau^2 - 0.208\,\tau^3. \tag{4.13}$$

In the pattern calculations of Fig. 4.5 [Mat.4.5] we have also computed the position and level of the first sidelobe as function of the illumination taper. Thus the first sidelobe level for uniform illumination is at -17.57 dB, decreasing to -24.64 dB for the full quadratic function.

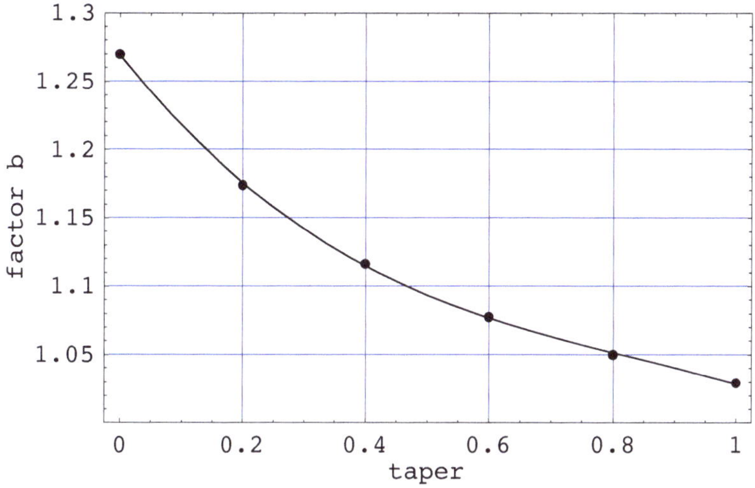

Fig. 4.6. The factor **b** in the HPBW formula as function of the taper τ.

We see a significant decrease in the level of the first sidelobe if the illumination becomes more tapered, which corresponds to a smaller value of τ. Next to the desire to minimise spillover radiation along the edge of the reflector, the lower sidelobe level is an important factor in choosing a tapered illumination. Fig. 4.7 illustrates the

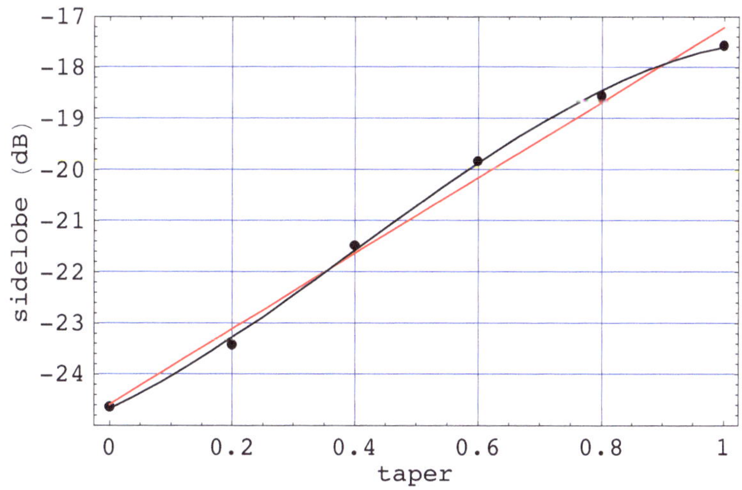

Fig. 4.7. The level of the first sidelobe in dB as function of the taper. The red line is the approximation of Eq. (4.14a), black is Eq. (4.14b).

sidelobe level in dB as function of the taper parameter τ [Mat.4.7]. To an accuracy of a few tenths of a decibel the relation can be approximated by the simple formula

$$S(\mathrm{dB}) = -24.6 + 7.37\ \tau\ . \tag{4.14a}$$

The calculated points are closely fit by the relation

$$\mathrm{S}\ (\mathrm{dB}) = -24.682 + 5.712\,\tau + 7.52\,\tau^2 - 6.156\,\tau^3\ . \tag{4.14b}$$

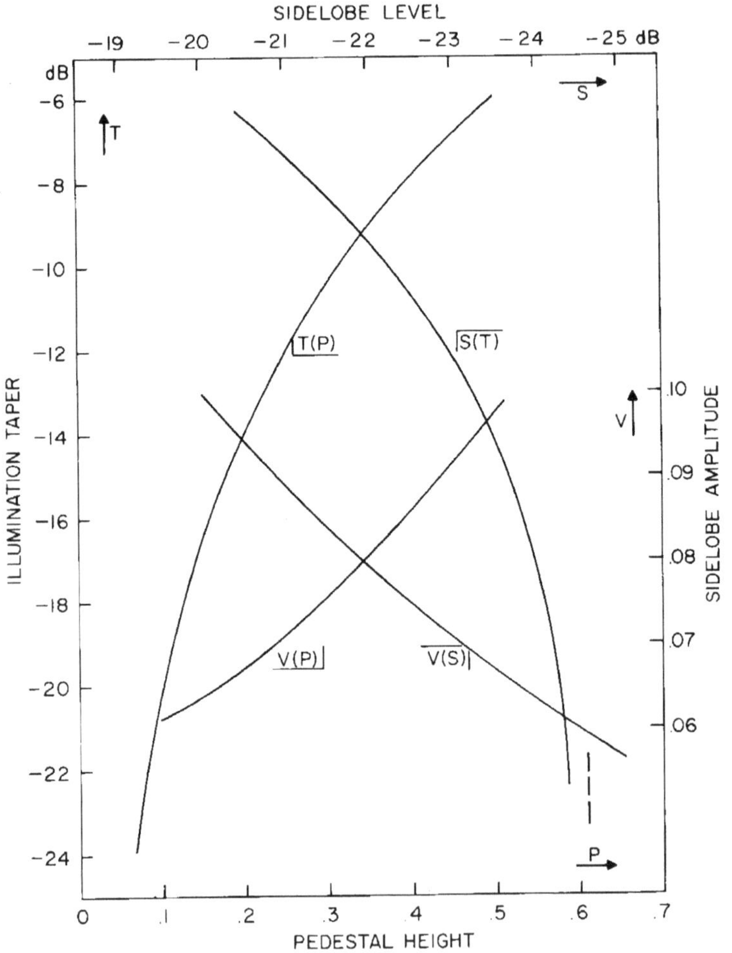

Fig. 4.8. The relation between illumination taper and level of first sidelobe. The short lines at the curve designation indicate the axes pertaining to the curve.

Some of the relations derived above are collected in a set of graphs in Fig 4.8. From these one can read the relations between the edge taper, expressed either in dB or in pedestal height value p, and the level of the first sidelobe in amplitude or dB level.

In Eq. (4.8) the illumination efficiency was given as function of the taper parameter τ. One could use this expression to derive a "tapered aperture diameter" d_{eff}, defined as the diameter of a uniformly illuminated reflector with the same effective area as the real reflector with tapered illumination (Mangum and Cheng, 1998). Thus d_{eff} is proportional to the square root of η_i, as given in Eq. (4.8). Using the effective diameter one could compute the expected change in HPBW from the simple relation that the HPBW is inversely proportional to the diameter of the reflector. The result gives values which are smaller than the HPBW, determined directly from the computed radiation pattern. *The concept of effective diameter seems to have only a limited usefulness in the analysis of the antenna behaviour.*

It is moreover not allowed to use the "tapered reflector radius" in the calculation of the components of the aperture efficiency. Based on the definition of aperture efficiency, its efficiency components must all be referenced to the full geometrical aperture area in order to arrive at the correct value of the overall aperture efficiency. Introducing a "tapered value" for some of those efficiency components is not allowed, because the illumination efficiency already has accounted for it. It is also worth remarking that in all our analyses we use the aperture plane as reference area, not the physical area of the curved reflector. (See also Chapter 5).

In the following sections we turn our attention to the influence of imperfections in the antenna on the radiation characteristics. The most important of these are a position of the feed displaced from the true focus and deviations of the reflector surface from the prescribed form. These are aspects of great practical value in the operation of the antenna or radio telescope.

4.3. Axial defocus

In Chapter 2 we described the geometrical relations of the paraboloid and presented the formulae for the path length differences caused by defocusing. Their influence on the characteristics of the radiation pattern will now be discussed. In general terms this is called the theory of *aberrations*. We will not give an exhaustive treatment (for this see e.g. Born and Wolf, 1980), but will derive the changes in beam parameters, like gain, beamwidth, sidelobe level, etc. First we treat the effects of *axial defocus*; that is the situation where the feed (or the subreflector in a Cassegrain system) is displaced along the paraboloid's axis from the true focal point.

From Ch. 2.2.2 we take Eq. (2.25) as a sufficiently accurate and convenient expression for the path length error due to axial defocus with the adjustment for the maximum phase error as discussed there. The maximum phase error at the aperture edge is thereby proportional to $(1 - \cos\psi_0)$. We shall demonstrate below that this approximation is fully warranted. This leads to the phase term over the aperture

$$\Delta\Phi(r) = \frac{2\pi\delta}{\lambda}(1 - \cos\psi_0)\, r^2 \equiv \beta r^2 . \qquad (4.15)$$

where ψ_0 is the opening half-angle of the paraboloid, dependent on the focal ratio through Eq. (2.10) and β replaces the term in front of r^2. It is obvious that axial defocus causes a phase error independent of the azimuthal aperture coordinate. Thus the integration over this coordinate can be performed, as done in Ch. 3.5. We now insert this phase function into the radiation integral of Eq. (4.9), assuming an amplitude illumination function $F(r)$ with rotational symmetry, to obtain

$$f(u) = \int_0^1 F(r)\, J_0(u\,r) \exp[-\,i\,\Delta\Phi(r)]\, r\, dr, \tag{4.16}$$

where $J_0(u\,r)$ is the Bessel function of the first kind and order zero, $u = (k\,d/2)\sin\theta$ with θ the angular polar coordinate of the beam. For the illumination function we select the usual quadratic on a pedestal form $F(r) = 1-(1-\tau)\,r^2$ with pedestal (taper) value τ.

With Eq. (4.15) and assuming uniform illumination Eq. (4.16) can be written as

$$f(u,\ \beta) = \int_0^1 J_0(u\,r) \exp[-\,i\,\beta\,r^2]\, r\, dr \;=\; \exp[-i\,\beta] \int_0^1 J_0(u\,r) \exp[i\,\beta\,(1-r^2)]\, r\, dr.$$

Separating the integral in its real and imaginary parts we obtain

$$U_1(u,\ \beta) \;=\; \beta \int_0^1 J_0(u\,r) \cos[\beta(1-r^2)]\, r\, dr,$$

$$U_2(u,\ \beta) \;=\; \beta \int_0^1 J_0(u\,r) \sin[\beta(1-r^2)]\, r\, dr,$$

and $$f(u,\ \beta)\, \exp[i\,\beta] \;=\; \tfrac{1}{\beta}\, [U_1(u,\ \beta) \;+\; U_2(u,\ \beta)].$$

The functions U_1 and U_2 are Lommel functions, already mentioned in Chapter 3.6, which can be expanded into the following series

$$U_1(u,\ \beta) \;=\; \frac{2\beta}{u}\, J_1(u) \;-\; \left(\frac{2\beta}{u}\right)^3 J_3(u) \;+\; \left(\frac{2\beta}{u}\right)^5 J_5(u) \;-\; \ldots$$

$$U_2(u,\ \beta) \;=\; \left(\frac{2\beta}{u}\right)^2 J_2(u) \;-\; \left(\frac{2\beta}{u}\right)^4 J_4(u) \;+\; \ldots$$

and the power pattern of the antenna can now be written as

$$G(u,\ \beta) \;=\; \frac{1}{\beta^2}\, \{\, {U_1}^2(u,\ \beta) \;+\; {U_2}^2(u,\ \beta)\, \}$$

$$= 4\left(\frac{J_1(u)}{u}\right)^2 \;-\; 16\,\beta^2 \left(\frac{2\,J_1(u)\,J_3(u)}{u^4} \;-\; \frac{{J_2}^2(u)}{u^4}\right) \;+\; \ldots \tag{4.17}$$

The first term is the beam function for the in-focus situation ($\beta = 0$) and uniform illumination, as given in Eq. (4.10) .

Inspecting the radiation integral for the axially defocused paraboloid we notice a strong similarity with the integral of the focused antenna in the Fresnel region, as discussed in Section 3.6. Obviously Eq. (3.37) is very similar to Eq. (4.16) as shown next, with some trivial change in notation,

$$f(u) = \int_0^1 F(r)\, J_0(u\, r) \exp\left(i\, k\, \tfrac{r^2}{2R}\right) r\, dr\,, \qquad \text{from Eq. (3. 37)}$$

and

$$f(u) = \int_0^1 F(r)\, J_0(u\, r) \exp[-i\, \Delta\Phi(r)]\; r\, dr, \qquad \text{from Eq. (4.16).}$$

The integrals are identical if $\Delta\Phi(r) = -\,k\, r^2 \,/\, 2\,R$, or $k\,\delta(1 - \cos\psi_0\,)\, r^2 = -k\, r^2 \,/\, 2\,R$.

To a first approximation $(1 - \cos\psi_0) = 2\,(1\,/\,2\,f)^2$ (see Eq. (2.13) and we obtain

$$\delta = -\, f^2 \,/\, R\,. \tag{4.18}$$

Thus, *to a first approximation*, we can mimic the farfield radiation pattern at a distance R from the aperture by shifting the feed outwards along the axis by an amount δ given by Eq. (4.18). This fact is used in the measurement of antenna radiation patterns on pattern ranges of a limited, finite length. It is routinely used to offset nearfield effects in the holographic measurement of reflector surfaces with the aid of a transmitter at finite distance, We shall discuss this more completely in the treatment of the holography method in Chapter 6.

4.3.1. Gain function with axial defocus

For the calculation of the gain decrease due to the defocus, we need only to calculate the on-axis values of $f(u)$, i.e. for $u = 0$, where $J_0(0) = 1$. The integral simplifies to the integral over the illumination amplitude function $F(r)$ times the complex exponential phase term of Eq. (4.16), which we separate in sine and cosine terms. Wc obtain, maintaining the normalisation term introduced earlier and writing $z = r^2$,

$$f_c(0,\ \beta) = \frac{2}{1+\tau} \int_0^1 (1 - (1 - \tau)\, z) \cos(\beta\ z)\ dz \tag{4.19a}$$

$$f_s(0,\ \beta) = \frac{2}{1+\tau} \int_0^1 (1 - (1 - \tau)\, z) \sin(\beta\ z)\ dz \tag{4.19b}$$

and the on axis gain function with defocus becomes

$$g(0,\ \beta) = f_c^{\,2} + f_s^{\,2}\,. \tag{4.20}$$

Instead of using the representation in a series of Bessel functions as sketched above, we evaluate this directly with the aid of *Mathematica* [Mat.4.8]. The resulting gain function with axial defocusing for arbitrary taper τ is obtained as

$$\begin{aligned} g(0,\beta) = \frac{4}{\beta^4\,(1+\tau)^2}\,&(2+\beta^2-4\,\tau+(2+\beta^2)\,\tau^2 \\ &-2\,(1+\tau\,(-2+\beta^2+\tau))\,\mathrm{Cos}\,(\beta)-2\,\beta\,(-1+\tau)^2\,\mathrm{Sin}\,(\beta))\,, \end{aligned} \tag{4.21}$$

where β is the maximum phase error at the aperture edge, as given above in Eq. (4.15). For the case of uniform illumination ($\tau = 1$) the function reduces to

$$g(0,\ \beta) = 2\,(1-\cos\beta)\,/\,\beta^2 = \left\{\frac{\sin(\beta/2)}{(\beta/2)}\right\}^2 \approx 1 - \beta^2\,/\,12, \tag{4.22}$$

For the fully tapered illumination (τ=0) we obtain the result

$$\begin{aligned} g(0,\ \beta) &= 4\,(2+\beta^2-2\cos\beta-2\,\beta\sin\beta)\,/\,\beta^4 \\ &= \left\{\frac{\sin(\beta/2)}{\beta/2}\right\}^4 + \frac{4}{\beta^2}\left\{\frac{\sin\beta}{\beta}-1\right\}^2 \approx 1-\beta^2\,/\,18 \end{aligned} \tag{4.23}$$

The simple quadratic approximations are reasonably good for values of $\beta < 1$.

If we take the gain function, based on the Lommel functions above (Eq. (4.17)) and let u->0, the beam maximum, we obtain

$$g(\beta,\ 0) = 1 - 16\,\beta^2\,\lim_{u>0}\left(\frac{2\,J_1(u)\,J_3(u)-J_2^{\,2}(u)}{u^4} + ..\right) = 1 - \frac{\beta^2}{12}\,.$$

This is identical with the result found above (Eq. (4.22)). In his paper on antenna tolerances Bracewell (1961) obtains the same result along an heuristic argument.

In the discussion of the geometry of aberrations in Chapter 2 we concluded that a better approximation for the path length changes due to axial defocus is given by Eq. (2.28), while a fully correct function is presented in Eq. (2.29). The differences in the gain loss for these three approximations are shown in Fig. 4.9, [Mat.4.9]. There is barely a difference between the curves for the exact phase function calculation (red curve), the good approximation of Eq. (2.28) (green) and the further approximation of Eq. (4.22) (blue).

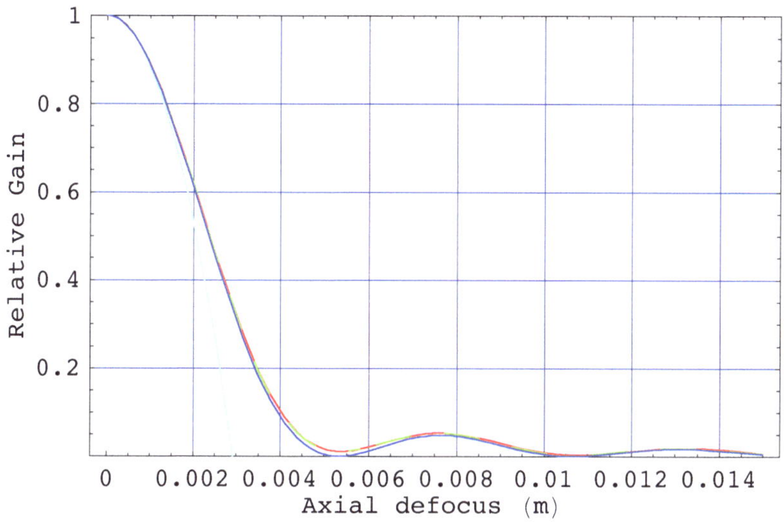

Fig. 4.9. Gain loss due to axial defocus of the feed from the primary focus for 3 mm wavelength and uniform illumination. The dashed red curve uses the exact phase error of Eq. (2.29), the green curve is for Eq. (2.28), essentially coinciding with the red, and the blue curve Eq. (4.22), barely separated from the other two. The cyan line is the quadratic approximation of Eq. (4.22) to the blue function, adequate for $\beta < 1.5$.

We return now to Eq. (4.21) which is the expression for the gain loss as function of axial defocus (β) for arbitrary taper value τ [Mat.4.10]. The function is plotted in Fig. 4.10, both on linear and logarithmic scale. It is clear that the gain loss is less with a highly tapered distribution and is also lacking the sharp minimum. This can be explained by the lesser destructive interference between the differently weighted phase errors of the inner and outer sections of the aperture.

The results indicate that it is important to maintain a proper axial focus. A defocus of the primary focus feed, or the subreflector in a Cassegrain system, of only half a wavelength causes already a gain loss of 30 percent. As we shall see later, the focal length for optimum gain is often dependent on the elevation angle of the antenna. A remotely controlled axial position system for the feed is thus required to adjust the focus to maximum gain dependent on the elevation angle. This feature is available in all modern antennas and radio telescopes.

4.3.2. Beamwidth and sidelobe variation with axial defocus

It is of interest to also investigate the change in beamwidth and sidelobe level with defocus. We expect to see a widening of the beam and an increase in the sidelobe level. For this we must integrate the full equation (4.16):

$$f(u) = \int_0^1 F(r)\, J_0(u\, r) \exp[-\, i\, \Delta\Phi(r)]\, r\, dr.$$

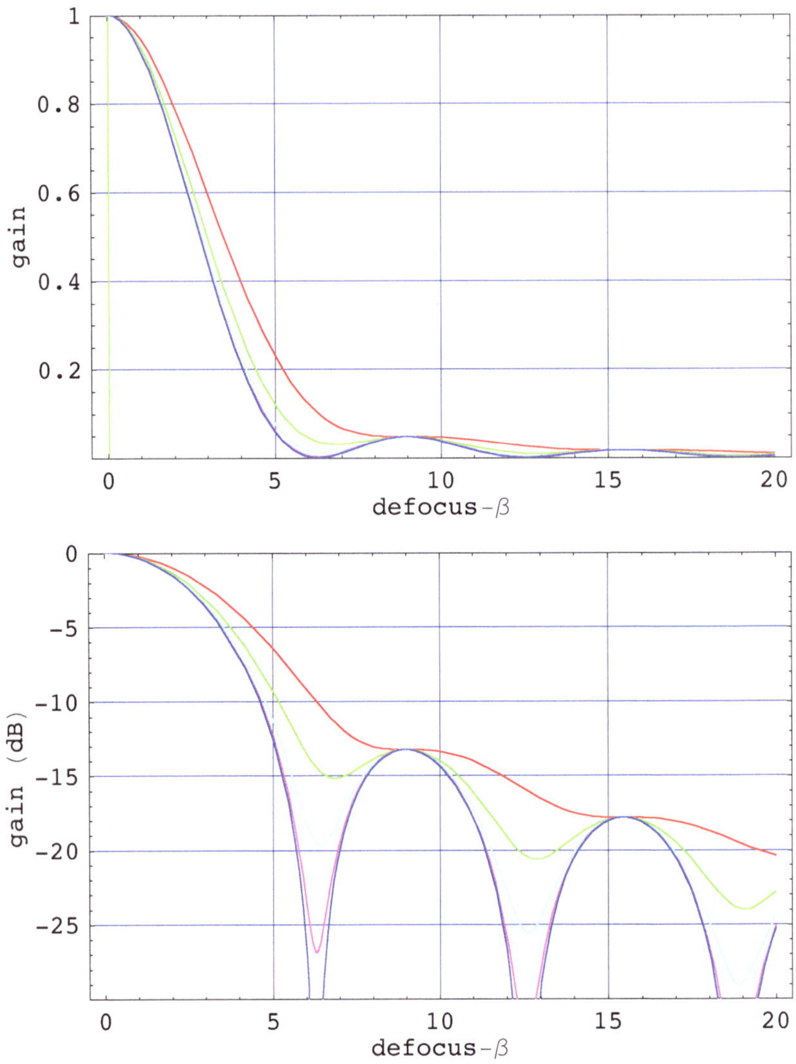

Fig. 4.10. Gain loss due to axial defocus for tapered aperture illuminations with τ = (0, step 0.25, 1), top to bottom. The abscissa shows the maximum phase error β, which can be changed to a defocus through Eq. (4.15).

In the following calculation we present a series of beam patterns for increasing values of the axial defocus in steps of 1 mm for a wavelength of 3 mm and the geometry of the ALMA antenna [Mat.4.11]. In this calculation we have used the simple quadratic function in r for the phase error over the aperture (Eq. (4.15)). We take the uniform illumination first, shown in Fig. 4.11.

The more practical case for an illumination function with -12 dB taper (τ = 0.25) is shown in the following figure (Fig. 4.12) [Mat.4.12]. Again we see that the influence of the defocus is "smoothed" in the case of tapered illumination. It is obvious that the

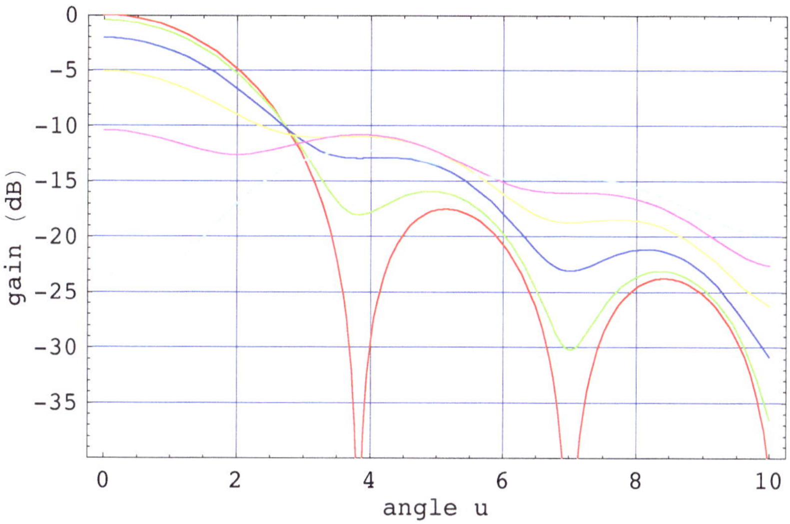

Fig. 4.11. A set of beam patterns for different values of axial defocus and a uniform illumination function.The defocus is in steps of 1 mm from 0 to 5 (red, green to cyan) and the wavelength is 3 mm.The upper curve (red) applies to the focused case.

sidelobe structure is quickly destroyed; already with a defocus of 2 mm (two-thirds of the wavelength) the sidelobe has "smeared" to a broad shoulder. We could have used the Lommel function representation, and its series development in Bessel functions to obtain this result. However, the direct integration can be performed without approximations within *Mathematica* and gives a quicker and more trustworthy result.

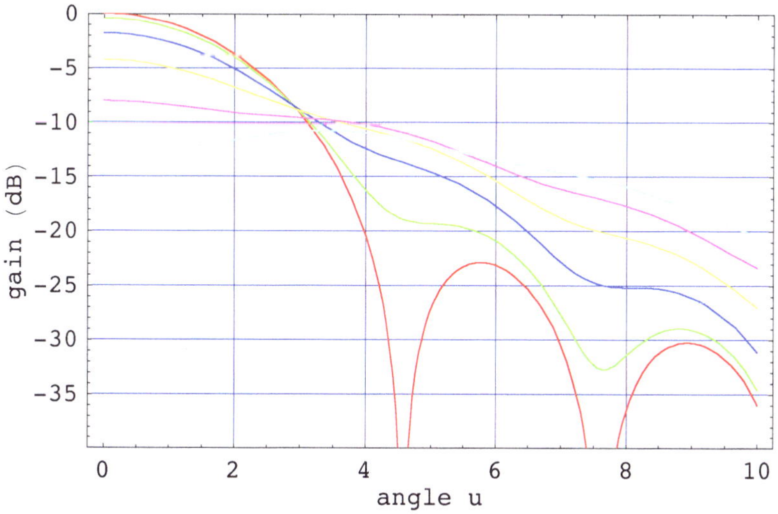

Fig. 4.12. A set of beam patterns for different values of axial defocus and a quadratic illumination function with edge taper of-12 dB.The defocus is in steps of 1 mm from 0 to 5 and the wavelength is 3 mm.The upper curve (red) applies to the focused case.

An analytical calculation of the change in beamwidth is tedious, if at all possible. Numerical computations indicate that the beamwidth increases slightly with defocus and is strongly dependent on the illumination taper. The relative increase is roughly proportional to β^2 and our calculations for a realistic -12dB illumination taper indicate the following relation (derived from Fig. 4.12):

$$\theta_A \;/\; \theta_{A0} \;=\; 1 + 0.01\,\beta^2\,, \tag{4.24}$$

The beam broadening is too small to be used as a reliable indicator of a possible axial defocus. As we saw above, both the decrease in gain and the level of the first side-lobe present more sensitive criteria. We shall discuss the methods of determining the best focal position in Chapter 5.

4.3.3. Depth of focus in prime focus and Cassegrain configuration

So far we have dealt with the movement of the feed near the primary focus of the paraboloidal reflector. The discussion is valid without any change for axial defocus of the feed in the secondary focus of a Cassegrain reflector configuration. We only have to consider the equivalent primary-fed paraboloid with a focal length of **m** times that of the real primary, where **m** is the magnification factor of the Cassegrain system, given by the expression (Eq. (2.18))

$$m \;=\; \tan\!\left(\tfrac{\Psi_0}{2}\right) \;/\; \tan\!\left(\tfrac{\Phi_0}{2}\right),$$

where Φ_0 is the opening half-angle of the secondary reflector from the Cassegrain (secondary) focus and Ψ_0 is the opening half-angle of the primary, as used before. Now, if we would use the approximation of Eq. (2.25) for the phase error due to defocus, it is immediately clear that an axial defocus in the secondary focus can be m^2 times as large as in the primary focus to cause the same phase error, and hence gain loss. As we have seen, a more correct approximation is given by Eq. (2.28) and from this we find that for an equal phase error over the aperture the ratio of the defocus in secondary δ_s to primary δ_p focus is given by

$$\frac{\delta_s}{\delta_p} \;=\; \frac{2\left(\frac{d}{4f}\right)^2}{1+\left(\frac{d}{4f}\right)^2} \Bigg/ \frac{2m^{-2}\left(\frac{d}{4f}\right)^2}{1+m^{-2}\left(\frac{d}{4f}\right)^2} \;=\; m^2\,\frac{1+\frac{1}{m^2}\left(\frac{d}{4f}\right)^2}{1+\left(\frac{d}{4f}\right)^2}\,. \tag{4.25}$$

Normally m $\gtrsim$ 10 and for a typical value of $f/d = 0.4$, the quotient in Eq. (4.25) has a value of about 0.7. The "depth of focus" factor of a Cassegrain system with respect to the primary is given by the above formula. In *Mathematica* it takes the form (numerical values valid for the ALMA antenna):

```
m = 20; F = f / d; F = 0.4;
dep_foc = m^2 (1 + m^-2 (4 F)^-2) / (1 + (4 F)^-2)
```

Out[105]= 287.921

Thus an axial focus error in the Cassegrain focus is about 300 times less severe than in the primary focus of the reflector.

In the Cassegrain configuration an axial shift of the secondary reflector δ_s from the primary focus causes a phase error over the primary reflector as with a primary feed shift plus a small phase error over the secondary from its displacement with respect to the feed. Thus the phase error is given by

$$\Delta\Phi = \frac{2\pi}{\lambda}\,\delta_s\,(1 - \cos\phi + 1 - \cos\psi) \approx \frac{2\pi}{\lambda}\,\delta_s\,(1 - \cos\psi), \qquad (4.26)$$

the final approximation being allowed, because in almost all cases the subtended angle of the subreflector is a few degrees and $\cos\phi \approx 1$. In other words, the treatment above for the primary feed can be used without significant loss of accuracy for the axial shift of the secondary reflector.

■ 4.4. Lateral defocus - Coma, Beam-Deviation-Factor

4.4.1. Off-axis beam function - Coma

We now turn to a discussion of the effects of a lateral defocus of the feed, or the secondary reflector, from the focal point. The geometry has been presented in Ch. 2.2.1. Note that in this case, contrary to axial defocus, the phase error will be dependent on the azimuthal aperture coordinate χ which complicates the analysis. We shall begin with the full phase error, but then limit our calculations to the beam characteristics in the plane of defocus, where the effects are of course most pronounced. From the discussion in Ch. 2.2.1 we repeat here the formula for the phase error over the aperture, under the assumption that $(\delta / f) << 1$ (Eq. (2.23),

$$\Delta\Phi(r, \chi) = \frac{2\pi}{\lambda}\,\delta\,\frac{r}{f}\left(\frac{1}{1+\left(\frac{r}{2f}\right)^2}\right)\cos\chi. \qquad (4.27)$$

We must now introduce this phase error in the radiation integral of the antenna, as given by Eq. (3.29),

$$f(\theta, \phi) = \int_0^1 \int_0^{2\pi} F(r, \chi) \exp\left[-i \left\{k \frac{d\,r}{2} \sin\theta \cos(\phi - \chi) + \Delta\Phi(r, \chi)\right\}\right] r\, d\,r\, d\,\chi. \tag{4.28}$$

We assume that the defocus does not have an effect on the amplitude distribution of the aperture field; only the phase function will be modified. This assumption is justified because we consider only small values of the defocus. By tilting the off-axis feed so as to direct the maximum of its radiation function towards the center of the aperture, we assure a symmetrical amplitude distribution as well as possible. The aperture integration integral of Eq. (4.28) has been normalised to a radius of one. To adjust the phase error to a unit aperture radius, we must multiply the r-terms in Eq. (4.27) by $d/2$. Writing $\Xi = 4f/d$, the radiation integral now becomes (ignoring constants before the integral sign)

$$f(\theta, \phi) = \int_0^1 \int_0^{2\pi} F(r) \exp\left[-i\,k\, \frac{d}{2}\, r \left\{ \sin\theta \cos(\chi - \phi) - \frac{\delta}{f}\, \frac{1}{1+\left(\frac{r}{\Xi}\right)^2} \cos\chi \right\}\right] r\, d\,r\, d\,\chi. \tag{4.29}$$

The integration over χ can be performed readily, provided we can obtain one term in χ. This can be achieved by the following substitution (Baars, 1970). We introduce the variable

$$K = \frac{\delta}{f} \left(\frac{1}{1+\left(\frac{r}{\Xi}\right)^2} \right)$$

and the term in curly brackets of Eq. (4.29) can now be written as

$$\sin\theta \cos(\chi - \phi) - K\cos\chi = (\sin\theta\ \cos\phi - K)\cos\chi + \sin\theta \sin\phi \sin\chi = \Gamma\cos\gamma \cos\chi + \Gamma\sin\gamma \sin\chi = \Gamma\cos(\gamma - \chi)\,,$$

where we have introduced the new variables Γ and γ, which have the form

$$\Gamma^2 = (\sin\theta\cos\phi - K)^2 + (\sin\theta\sin\phi)^2 = \sin^2\theta - 2K\sin\theta\cos\phi + K^2 \tag{4.30a}$$

and

$$\tan\gamma = \frac{\sin\theta\ \sin\phi}{\sin\theta\ \cos\phi - K}\,. \tag{4.30b}$$

With these substitutions the integral of Eq. (4.29) becomes

$$f(\theta,\ \phi) = \int_0^1 \int_0^{2\pi} F(r) \exp\left[-i\, k\, \tfrac{d}{2}\, r\ \Gamma(\theta,\ r,\ \phi)\ \cos(\gamma - \chi)\right] r\, d r\, d \chi, \tag{4.31}$$

in which the integration over χ can now be performed to yield

$$f(\theta,\ \phi) = 2\pi \int_0^1 F(r)\ J_0\left[k\, \tfrac{d}{2}\ r\, \Gamma(r,\ \theta,\ \phi)\right] r\, d r. \tag{4.32}$$

This radiation pattern will not be rotationally symmetric in ϕ because Γ is dependent on ϕ. For the geometry, refer to Fig. 3.1 and recall that θ is the polar angle of the radiation pattern, while ϕ is the azimuthal angle of the pattern. In the most interesting plane, the one in which the feed has been displaced (often called the plane of scan), we take $\phi = 0$ and hence $\Gamma = \sin\theta - K$. The K-term causes a beam shift in the plane of feed movement and the coma lobe due to the higher order term in r. If the effective focal length is large, as in a Cassegrain configuration where typically the magnification is of order 10, the quadratic r-term in K can be neglected and the beam squint is simply $-\delta / f$. For a prime focus reflector the full equation for K must be preserved in the integral. With $\Xi = 4\, f / d$, as introduced above, and using, as before, $u = \frac{k\,d}{2} \sin\theta$ the integral in the plane of the defocus ($\phi = 0$) takes the final form (ignoring irrelevant constants)

$$f(u) = 2 \int_0^1 F(r)\, J_0\left[\, r\left(u - \frac{2\, k\, \delta}{\Xi}\ \frac{1}{1+\left(\frac{r}{\Xi}\right)^2}\right)\right] r\, d r. \tag{4.33}$$

This formula can be written as a *Mathematica* expression for the computation of off axis radiation patterns [Mat.4.13]. An example for the ALMA antenna follows. Note that the units for d, δ and λ must be the same. The feed is displaced by one-third of a wavelength with each step. From Fig. 4.13 we see that the gain loss is small, less than 0.5 dB, but the sidelobe towards the reflector axis (the Coma-lobe) increases strongly to -13 dB for a lateral defocus of only one wavelength. However, the sidelobes on the other side of the beam become very weak. This is a good criterion to determine the correct lateral focus in practice.

We illustrate the general appearance of these off-axis beams with a few more figures, all derived with the aid of *Mathematica* [Mat.4.14]. On the right side of Fig. 4.14 a contour plot of the beam with the feed laterally displaced by one wavelength is shown. The "waning moon"-like coma lobe is clearly visible to the left of the bright central main beam. A 3-dimensional plot is shown in the left part of the figure. Similar calculations are presented next for the case of -12 dB illumination taper (a widely used value in practice) [Mat.4.15] The plot of Fig. 4.15 show the results. The gain loss is somewhat less and the coma lobe is not as strong. This is again a result of the decrease in constructive or destructive interference of the reflected, "secondary" wavefronts caused by the difference in amplitude of these waves depending on the location of their origin on the reflector.

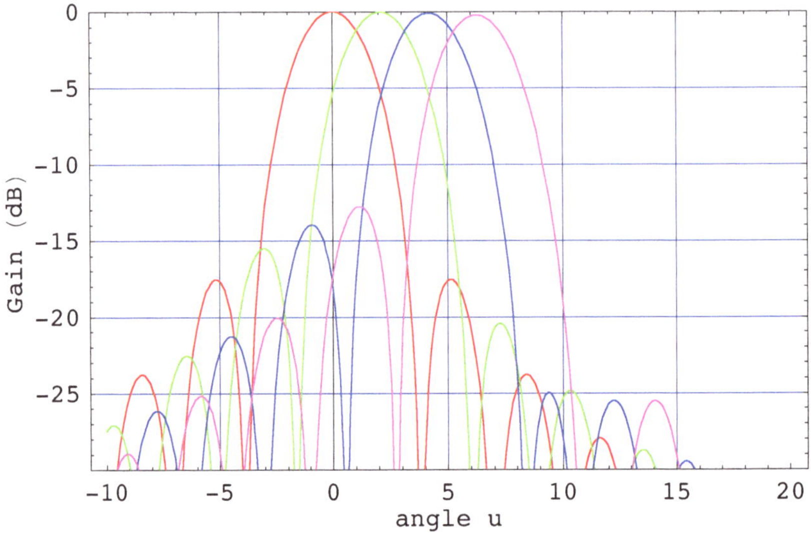

Fig. 4.13. Radiation patterns with uniform illumination and for lateral feed displacement of up to one wavelength in three equal steps. Note the strong sidelobe on one side -the Coma lobe. The gain loss is relatively small.

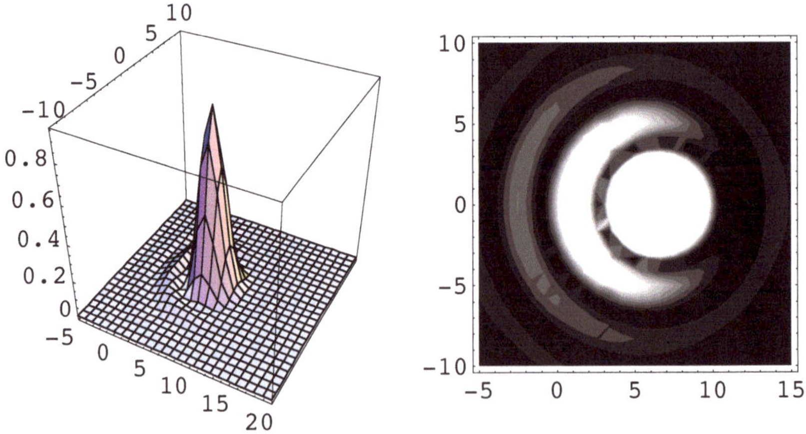

Fig. 4.14. A 3-dimensional and a contour plot of the antenna beam with the feed displaced laterally by one wavelength. Note the coma lobe in the form of the "waning moon" crescent on the left of the main beam. The brightness of the main beam has been truncated to make the coma lobe more visible.

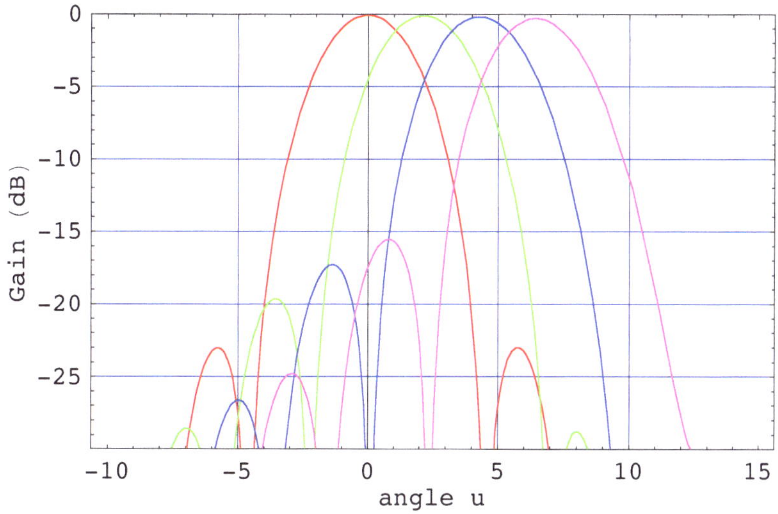

Fig. 4.15. Radiation patterns for the same lateral feed displacement, but in the case of an illumination function with -12 dB taper.

4.4.2. Gain and sidelobe level for off-axis beam

From these calculations we can construct tables or graphs of the off-axis behaviour of the antenna. Important quantities are the gain loss and the coma lobe level. These cannot be computed in a straightforward, analytical way. However, the *Mathematica* expressions **FindRoot** and **FindMaximum** offer an easy way to find the numerical values. Fig..4.16 depicts these quantities, as deduced from the calculations, for three illumination functions: uniform, -12 dB taper, quadratic taper (zero at edge) and for the ALMA antenna geometry with $f/d = 0.4$. The curves are polynomial fits to the computed points (black) at p = 0 to 5.The results of the fits are given in Table 4.1.

Note that the parameter p is reckoned in half-power beam widths (HPBW) off-axis. The physical shift of the feed (or subreflector) needed for such a beam shift is dependent on the illumination taper and the primary f/d -ratio. We can write for the feed displacement δ_l per HPBW beam shift

$$\delta_l \,/\, \text{HPBW} = \frac{b}{\text{BDF}} \frac{f}{d} \lambda, \tag{4.34}$$

where BDF is the beam deviation factor, to be discussed in the next section. The general expression for the HPBW is $\Theta_A = b\,(\lambda/d)$ (in radian), where b depends on the illumination function. For a "quadratic on a pedestal" illumination function with taper τ (Eq. 4.7), a useful approximation for b as function of τ is given in Eq. (4.13). In case of the ALMA antenna we have, for a -12 dB tapered illumination ($\tau = 0.25$), BDF = 0.82, $b =$ 1.165 and we find for the lateral feed shift per HPBW beam shift

$$\delta_l \,/\, \text{HPBW} = (p = 1) = 0.57\,\lambda\,.$$

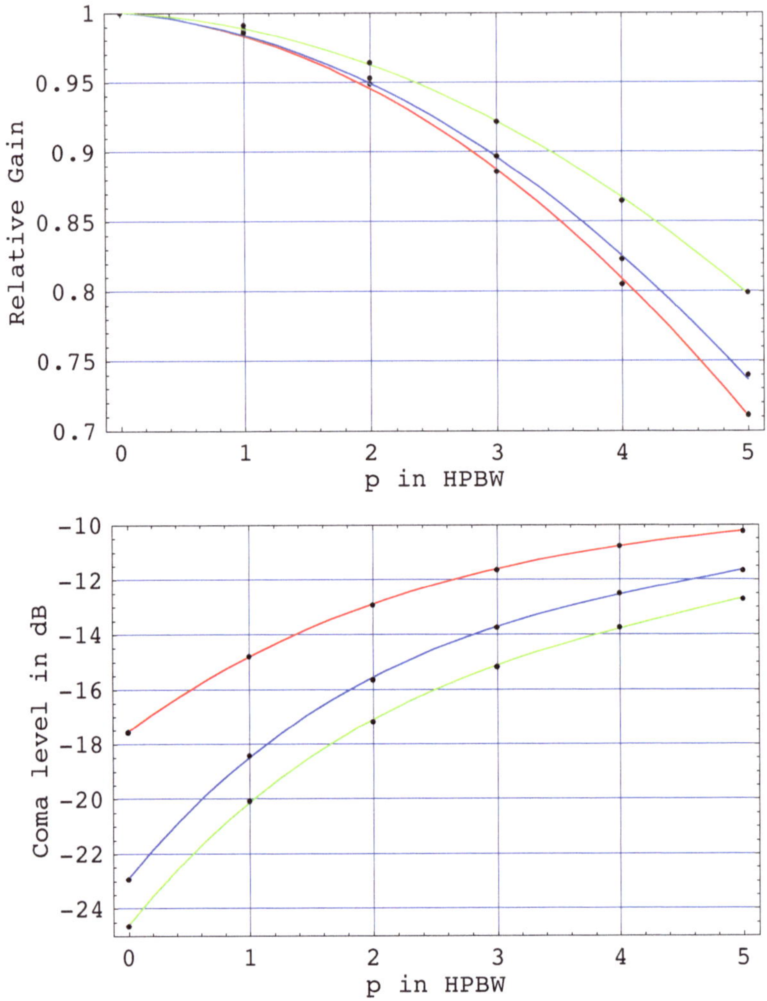

Fig. 4.16. Gain loss and Coma-lobe level as function of lateral defocus for uniform (red), -12 dB (blue) and full (green) illumination taper

Table 4.1. Gain loss and Coma lobe with lateral defocus

Illumination	Gain loss function	Coma lobe level (dB)
Uniform	$1 - 0.0070\,p - 0.0102\,p^2$	$-17.6 + 3.2\,p - 0.50\,p^2 + 0.03\,p^3$
Taper -12 dB	$1 - 0.0073\,p - 0.0091\,p^2$	$-23.0 + 5.3\,p - 0.95\,p^2 + 0.07\,p^3$
Quadratic	$1 - 0.0045\,p - 0.0072\,p^2$	$-24.6 + 5.3\,p - 0.93\,p^2 + 0.07\,p^3$

The calculations presented here are all for a prime focus geometry and a movement of the feed from the focal point. The results are valid to good accuracy for the displacement of the secondary reflector from the focus. As remarked before, the situation in the secondary focus is far less critical because of the large magnification factor m of the Cassegrain system. A lateral feed shift is m times less critical, while the axial "depth of focus" is approximately 0.5 m^2 less critical than in the primary focus, as illustrated earlier. Also, a rotation of the secondary causes a beam shift which is about m times smaller than the actual secondary rotation (see Ch. 5.5.1).

4.4.3. Beam Deviation Factor (BDF)

We turn now to an important quantity, related to lateral defocus; the *beam deviation factor*. As we have seen, a lateral movement of the feed or subreflector from the focal point causes a shift of the beam away from the axis; i.e. the antenna exhibits a pointing error. The beam deviation factor (BDF) is defined as the ratio of the pointing shift to the amount of defocus (in angular measure). It is a number smaller than one and it is caused by the curved shape of the reflector. Qualitatively it is easy to see that the BDF will be smaller for a "deeper" reflector, i.e. one with a smaller f/d -ratio. The BDF is also dependent on the illumination function. The functional form of the BDF has been given by Ruze (1965). Following Ruze, we find the angular position of the peak of the beam where the squared phase error, weighted by the illumination function, is minimum. This condition can be written as (see Eq. (4.29)

$$\frac{\partial}{\partial u} \int_0^1 F(r)\,\{k\, r\,(\sin\theta - K)\}^2\, r\, d r = 0,$$

and carrying out the differentiation, we find the beam angle of maximum intensity u_b, which we can compare to the feed squint angle $u_f = \delta / f$. Using $\Xi = 4\, f/d$, as before, the resulting expression is

$$\mathrm{BDF} = \frac{u_b}{u_f} = \int_0^1 \frac{F(r)\, r^3}{1+(\frac{r}{\Xi})^2}\, d r \bigg/ \int_0^1 F(r)\, r^3\, d r. \qquad (4.35a)$$

The *Mathematica* expressions [Mat.4.16-18] perform the integrations and produce the plots of Figure 4.17a for the BDF as function of the reflector focal ratio for a number of illumination functions and Fig. 4.17b for BDF as function of taper for several focal ratios. The 3-D plot of [Mat 4.18] is not shown here. The formula for the BDF, resulting from the integration of Eq. (4.35a) can be written as

$$\mathrm{BDF} = \frac{1}{1+2\tau}\,\{48\,\mathrm{f}^2\,[1 - 32\,\mathrm{f}^2\,(-1+\tau) + \tau + 32\,\mathrm{f}^2\,(-1 + 16\,\mathrm{f}^2\,(-1+\tau))\,(-\mathrm{Log}\,(\mathrm{f}^2) + \mathrm{Log}\,(\tfrac{1}{16} + \mathrm{f}^2))]\}\,. \qquad (4.35\,b)$$

Here f is the focal ratio (f=f/d), τ is the taper parameter (Eq. (4.7)) and the natural logarithm should be taken.

As the curves indicate, the BDF can change by close to 10 percent depending on the illumination taper for "standard" focal ratios between 0.3 and 0.5. For multi-feed systems, it is important to establish the taper accurately in order to calculate the angular distance between the beams.

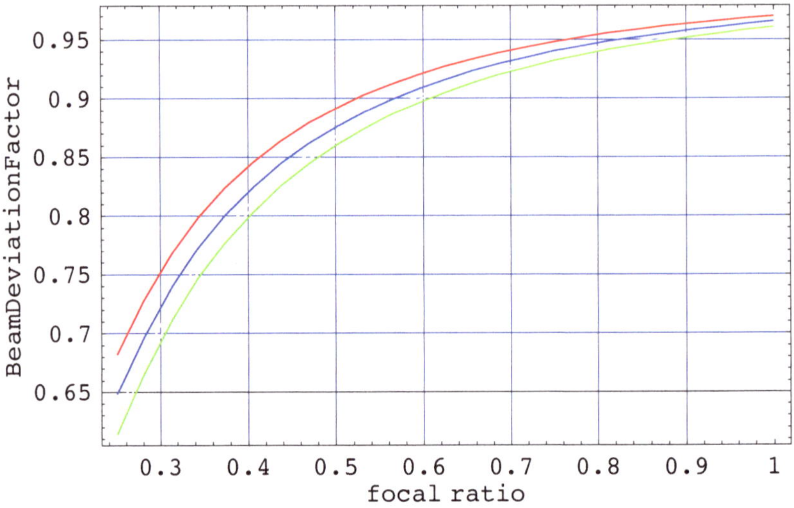

Fig. 4.17a. The BDF for fully tapered (red) and uniform (green) illumination as function of the focal ratio of the reflector. The middle curve (blue) is for a value of τ = 0.25 (-12 dB) taper.

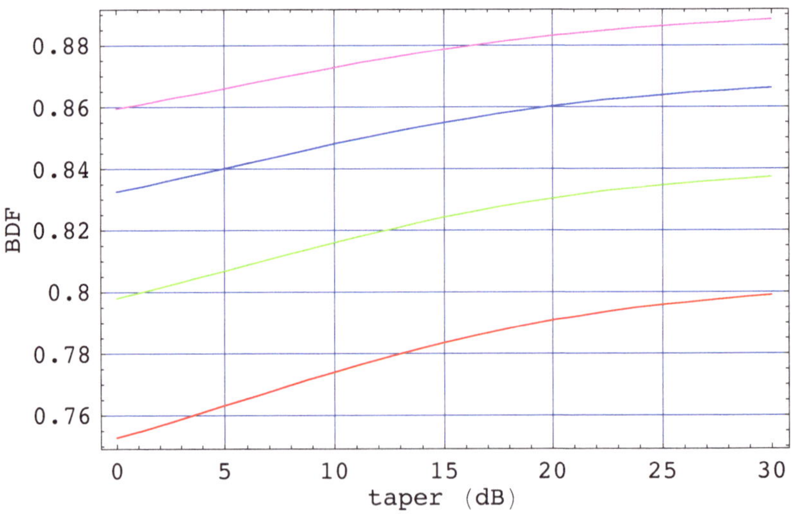

Fig. 4.17b. The beam deviation factor (BDF) as function of taper for f/D=0.35-red, 0.4-green, 0.45-blue, 0.5-magenta.

■ 4.5. Aperture blocking

In the common cylindrically symmetric layout of the Cassegrain antenna, the secondary reflector and its support structure (usually a four legged structure, called quadripod) cast a shadow of the incoming wavefront onto the aperture. This is normally called **aperture blocking** or **blockage**. In this section we present the formulae for the calculation of the blocked area and the resulting loss in aperture efficiency. Fig. 4.18 depicts the geometry and defines several variables. We assume a diameter of the central hole in the primary reflector not larger than the diameter of the secondary. This is generally the case.

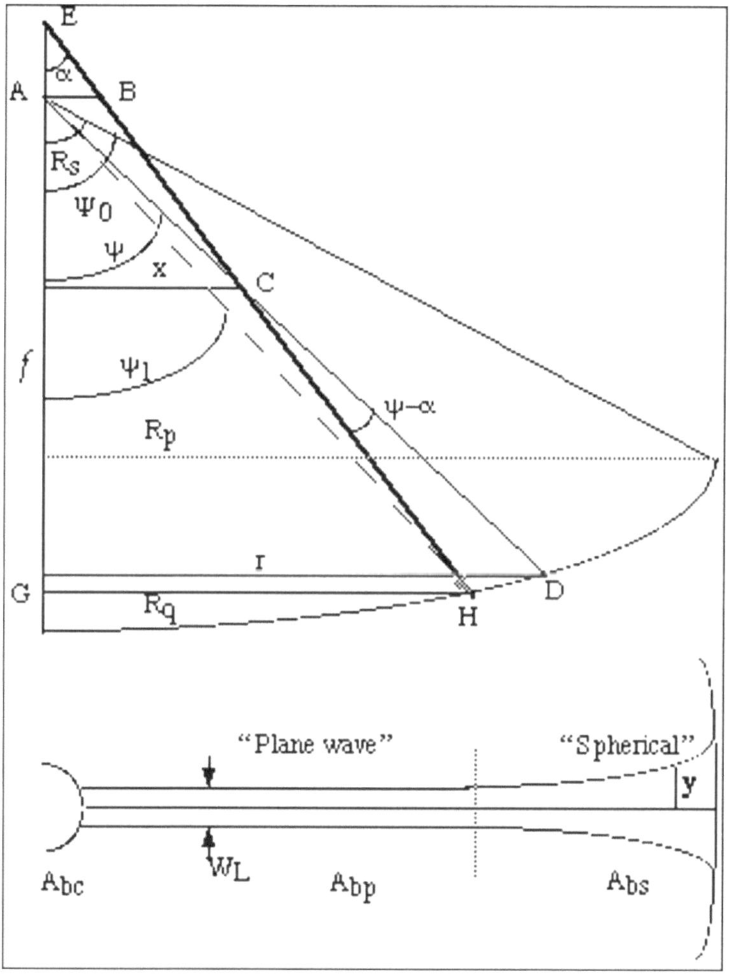

Fig. 4.18. Geometry of aperture blocking by feed struts and central obscuration.

The shadow on the main reflector, or the aperture, comprises three parts:

1. the central (circular) shadow of the subreflector,

2. "plane wave" shadow of the obstructed area of the infalling wavefront. It is the projection of the quadripod onto the aperture of the main reflector.

3. "spherical wave" shadow, which is caused by the obstruction of the reflected (spherical) waves from the outside section of the reflector on their way to the primary focus. This is the projection of the quadripod onto the aperture as seen from the focus of the primary reflector. It takes the form of approximately a trapezoid growing in width from the point where the legs cross the primary reflector towards the edge of the aperture. Inspection of Fig. 4.18 shows that the spherical shadow can easily be the major component of the aperture blocking. To minimize the spherical component, the legs should be directed as far to the outside of the aperture as possible. Often this is a compromise between structural and electro-magnetic considerations. Also, the spherical blocking will be smaller if the quadripod passes farther from the edge of the subreflector. This is the reason for the use of a curved quadripod as for instance in the ALMA antenna, designed by Vertex.

In the following sections we shall define the variables and set up the equations for the blocking components. We follow to a large extent the treatment of E. Maanders, as presented in his doctoral thesis (Maanders, 1975). The paper by Ruze (1968) provides a general introduction to the subject.

4.5.1. The variables and equations

We commence with the definition of the variables - see Fig. 4.18.

R_p - radius of primary — D - diameter of primary reflector

R_s - radius of subreflector — d - diameter of subreflector

f - focal length of main reflector — $f^* = f/D$ - focal ratio of primary

W_l - width of leg — n - number of legs

A_b - total blocked area; — A_{bs} - spherical wave component

A_{bp} - plane wave component, — A_{bc} - central component of blocked areas

$\Psi_0 = 2\arctan(D/4f)$ - opening half angle of primary

$\psi_1 = 2\arctan(R_q/2f)$ - opening angle to leg penetration point

R_q - radius at point of penetration of quadripod through main reflector

α - angle of leg with respect to main reflector axis

Here follow the equations for the different blocking components, assuming *uniform illumination*:

1. Central obstruction due to subreflector

$$A_{bc} = \pi R_s^2 \tag{4.36a}$$

2. Plane wave shadow area

$$A_{bp} = n\, W_l\, (R_q - R_s\,) \tag{4.36b}$$

3. Calculating the spherical wave shadow area is somewhat more involved. With reference to Fig. 4.18, consider a ray from the focus (A) under an angle ψ with respect to the axis, which is blocked at the point C by the leg. The ray would hit the reflector surface at point D at radius r. The width of the shadow at this radius is called y (see lower part of Fig. 4.18). Now it is easy to show that

$$\frac{y}{W_L/2} = \frac{r}{x} = \frac{r}{\mathrm{AC}\sin\phi}\,.$$

We apply the sine-rule to the triangle ABC to obtain

$$\frac{\mathrm{AB}}{\sin(\psi-\alpha)} = \frac{\mathrm{AC}}{\sin(\pi/2+\alpha)} = \frac{\mathrm{AC}}{\cos\alpha}\,.$$

Thus we get

$$y = \frac{W_F}{2}\; r\; \frac{\sin(\psi-\alpha)}{\mathrm{AB}\cos\alpha\sin\psi} = \frac{W_L\, r}{2\,\mathrm{AB}}\,(1-\tan\alpha\cot\psi)\,.$$

Using some trigonometric relations and the fact that $\tan\frac{\psi}{2} = \frac{r}{2f}$ (Eq. 2.10a), we obtain

$$y = \frac{W_F}{2\,\mathrm{AB}}\left\{r - f\tan\alpha + \frac{r^2}{4f}\tan\alpha\right\}.$$

The length of AB is found as follows. Using the relation $\tan\psi_1 = R_q\,/\,\mathrm{AG}$ (Fig. 4.18) we find

$$\tan\alpha = \frac{R_q-\mathrm{AB}}{\mathrm{AG}} = \frac{R_q-\mathrm{AB}}{R_q}\,\tan\psi_1\,,$$

from which follows

$$\mathrm{AB} = R_q\,[1-\tan(\alpha)/\tan(\psi_1)].$$

To obtain the spherical blocking we must now integrate y with respect to r from R_q to R_p with due regard for the aperture illumination function. For a uniform illumination the spherical blocking term is

$$A_{bs} = n\int_{R_q}^{R_p} 2\,y\,dr$$

with the result:

$$A_{bs} = \frac{n\,W_l}{\mathrm{AB}}\left[\frac{R_p^{\,2}-R_q^{\,2}}{2} - (R_p-R_q)\,f\,\tan(\alpha) + \frac{\tan(\alpha)}{12\,f}\,(R_p^{\,3}-R_q^{\,3})\right]. \tag{4.36c}$$

A somewhat simpler approach approximates the curved shadow boundary with straight lines and results in the following *alternative formula for the spherical blocking*,, leading to only very small changes (~1%) in the numerical outcome of the spherical blocking term.

$$A_{\mathrm{bs}}(\mathrm{alt}) = \frac{n}{2}\, W_l(R_p - R_q)\left[1 + \frac{R_p}{\sin\Psi_0} \Big/ \left(\frac{\sin\alpha}{\sin(\Psi_0-\alpha)}\left(\frac{R_q{}^2}{4f} + \frac{R_q}{\tan\alpha} - f\right)\right)\right]. \tag{4.36d}$$

These formulae are derived from the geometry only and hence are valid for a uniform illumination of the aperture.

For a *tapered illumination function* we assume the usual quadratic on a pedestal illumination function of Eq. (4.7) with pedestal height τ.

1a. The influence on the small, central obscuration by the subreflector can be ignored.

For the plane wave and spherical wave components we obtain the following **correction terms** to be **subtracted** from Eqs. (4.36b) and (4.36c) or (4.36d) above, respectively:

2a. Correction to the plane wave blocking part

$$A_{\mathrm{bp}}(t) = \frac{n\,W_l\,(1-\tau)}{3\,R_p{}^2}\,(R_q{}^3 - R_s{}^3). \tag{4.36e}$$

3a. Correction to the spherical wave blocking component

$$A_{\mathrm{bs}}(t) = \frac{n\,W_l(1-\tau)}{\mathrm{AB}{*}R_p{}^2}\left[\frac{R_p{}^4-R_q{}^4}{4} - f\tan(\alpha)\,\frac{R_p{}^3-R_q{}^3}{3} + \frac{\tan(\alpha)}{20\,f}\,(R_p{}^5 - R_q{}^5)\right]. \tag{4.36f}$$

Note that the effect of the blocking is expressed as a shadow, projected onto the *aperture* of the reflector. Hence, the blocking percentage is obtained by dividing this number by the area of the aperture A_r, not the physical area of the curved reflector.

The *Mathematica* expressions [Mat.4.19-21] for the computation of the blocking, together with the necessary input parameters, are given in Section 4.7. The numerical values there are taken from one of the ALMA prototype antennas. As with all routines, the reader might replace the parameters by those of his own antenna.

4.5.2. Gain loss and sidelobe level increase due to blockage

The loss of gain caused by the blocking, which we call here the **blocking efficiency** η_b, is given by the following expression:

$$\eta_b = (1 - A_b / A_r)^2 \approx 1 - 2(A_b / A_r), \tag{4.37}$$

the approximation being accurate to 1 percent or better if $A_b / A_r < 0.1$, which will normally be the case. The "double" effect of the blocking can be understood physically by considering that the blocking causes:

a) a decrease in the reflector area exposed to the incoming wavefront, and

b) a reduction of the incoming energy available for reflection to the focus.

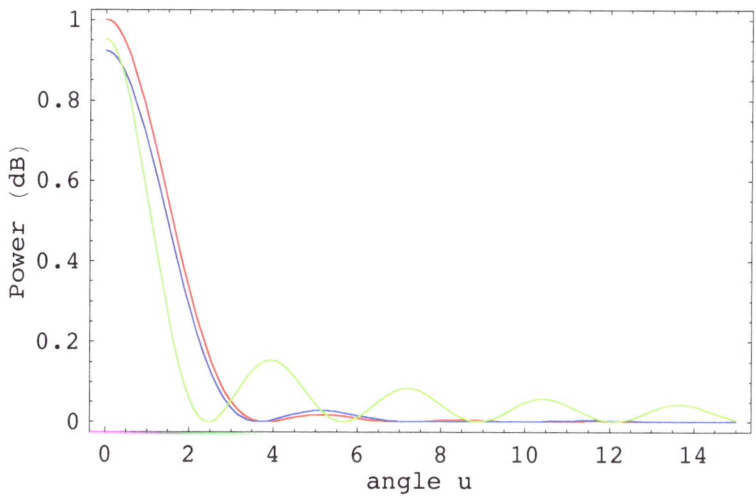

Fig. 4.19. Power patterns of a uniformly illuminated full area (red), central obscuration of 20% width (blue) and an outer annulus of 5 percent width (green), the last scaled up by a factor of 25.

A further result of the blocking is an increase of the sidelobe level due to the discontinuous aperture distribution and the scattering of the incoming wavefront by the blocking structures. A central obscuration with a diameter of 10% of the aperture diameter increases the first sidelobe by about 1 dB. This can be illustrated nicely by the realisation that the radiation function is now that of an annular aperture. In the following example (Fig. 4.19) [Mat.4.22] we compare the fields of a uniformly illuminated full aperture, one with a central obscuration of 20 percent of the diameter and an extreme case with a non-obscured annulus of only the outer 5 percent of the radius. We use in the first case Eq. (3.30) with $A(a) = 1$ and in the other the same expression, but with the lower integration value equal to 0.2 and 0.95, respectively. Clearly the half-power width decreases with increasing obscuration and the (first) sidelobe increases. The pattern of the very narrow annulus has very strong sidelobes with a much lower central peak amplitude (the plot in the figure has been multiplied by a factor 25 with respect to the others). When the width of the annulus tends to zero, we obtain the Bessel function of order zero (see e.g. Born and Wolf, Ch. 8.6.2).

A particularly interesting effect is the scattering of the obliquely infalling wavefront on the quadripod. The mathematical treatment is rather involved and has been presented by Rusch et al.(1982). It causes a *scattering cone* with a top angle determined by the angle α, i.e. typically some 40 degrees from boresight. The intensity of the "cone" is dependent on the width of the quadripod and is strongest if the width is of the order of the wavelength. Near the beam axis one sees an increased sidelobe level along the projection of the quadripod. An experimental example of these features is presented in Chapter 6.2.

■ 4.6. Reflector shape deviations - "surface tolerance theory"

4.6.1. Random surface deviation

Every antenna will suffer from fabrication errors and from deformations caused by gravity, wind and thermal effects. Often these will result in more or less randomly distributed deviations of the reflector surface from the theoretical shape. The influence of such random errors has been treated by Ruze (1952, 1966) and often these errors are called "Ruze error". It should be understood that the analysis of Ruze is strictly valid only for errors which are small with respect to the wavelength, are randomly distributed over the reflector with mean-square value $<\epsilon^2>$ and exhibit a "correlation length" $\boldsymbol{c}$ which is much smaller than the reflector diameter and much larger than the wavelength. In practice, it has been determined that these restrictions need not be satisfied rigorously in order to give satisfactory results. The physical effect of the random deviations is to remove power from the main beam and distribute this in a wide "scatter pattern" or "error beam" with a half power width proportional to the wavelength divided by the correlation length. Thus the total antenna beam can be written as the sum of the diffraction pattern and the error beam in the following form:

$$g(\theta) = g_D(\theta) + g_E(\theta,\ \epsilon) \tag{4.38}$$

We denote σ the root-mean-square phase fluctuation over the aperture, expressed by

$$\sigma = \frac{4\pi}{\lambda}\sqrt{<\epsilon>^2} \equiv \frac{4\pi\epsilon}{\lambda}. \tag{4.39}$$

For $\sigma = 1$ we have $\epsilon = \lambda/12.5$, which is a practical shortest wavelength for useful antenna operation, as we shall see below.

The error pattern can be written in closed form (due to Scheffler, 1962), rather than the infinite series of Ruze's original treatment, as

$$g_E(\theta) = \frac{1}{\eta_{A0}} \left(\frac{c}{d}\right)^2 \{\exp\sigma^2 - 1\} \exp\Big[-\Big(\frac{\pi c \theta}{2\lambda}\Big)^2\Big], \qquad \sigma^2 <= 1\,, \tag{4.40a}$$

$$g_E(\theta) = \frac{1}{\eta_{A0}} \left(\frac{c}{d}\right)^2 \{\exp\sigma^2 - 1\} \frac{1}{\sigma^2} \exp\Big[-\Big(\frac{\pi c \theta}{2\lambda\sigma}\Big)^2\Big], \quad \sigma^2 > 1. \tag{4.40b}$$

Here η_{A0} is the aperture efficiency of the perfect reflector (or $\lambda \rightarrow \infty$), θ the angular coordinate of the rotationally symmetric error pattern, and c and d the correlation length and reflector diameter, respectively, with c<<d.

If we integrate Eq. (4.38) over all space ($4\,\pi$ steradians) we obtain the antenna solid angle as (see Ch. 5)

$$\Omega_A = \Omega_0 + \Omega_0 \{\exp\sigma^2 - 1\} = \Omega_0 \exp\sigma^2, \tag{4.41}$$

where Ω_0 is the antenna solid angle of the perfect reflector. This increase of the antenna solid angle means an inversely proportional decrease in the aperture efficiency. The latter is proportional to the peak level of Eq. (4.38). From this expression and Eq. (4.40a) we obtain for the aperture efficiency η_A

$$\eta_A = \eta_{A0} \exp(-\sigma^2)\,(g_D(0) + g_E(0)\,).$$

Thus we find for the relative aperture efficiency caused by the random errors (see Ruze (1966) and Baars (1973)),

$$\eta_e = \frac{\eta_A}{\eta_{A0}} = \exp(-\sigma^2) + \frac{1}{\eta_{A0}} \left(\frac{c}{d}\right)^2 \{1 - \exp(-\sigma^2)\}, \tag{4.42}$$

The actual surface deviations ϵ_n are mostly calculated or measured *normally* to the reflector surface. However, the phase error is proportional to the pathlength change and hence ϵ is the component of this deviation ϵ_n parallel to the antenna axis (Fig. 4.20). The relationship between the two quantities is dependent on the focal ratio and is given by

$$\epsilon = \epsilon_n \cos\left(\frac{\psi}{2}\right) \equiv \frac{\epsilon_n}{\sqrt{1+(r/2\,f)^2}}\,. \tag{4.43}$$

Thus the errors towards the edge of the reflector have a smaller influence on the overall rms error. In practice, in order to calculate the expected gain loss caused by

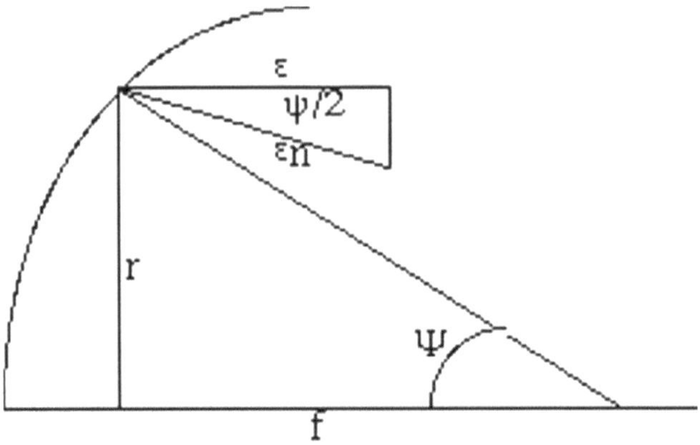

Fig. 4.20. Illustration of the relation between normal surface error ϵ_n and the axial component ϵ. The induced phase error is proportional to the latter.

the random surface deviations, one can also weigh the measured errors by the illumination function over the aperture, again diminishing the effect of the outside regions of the reflector (see e.g. Greve and Hooghoudt, 1981).

The first term in Eq. (4.42) is the usual form of the "Ruze" loss caused by small random surface errors. The second term is the maximum level of the "error beam", which is both dependent on the correlation length and the phase error. Normally, the second term is negligible for the resulting loss in antenna efficiency, but for relatively large phase errors and large correlation length it can be significant. In any case, it is of considerable significance for the observation of extended sources, because the error beam will be "filled" by the extended structure and lead to overestimation of the source's intensity. The peak level of the error beam with respect to the peak of the main beam is

$$\frac{g_E(0)}{g_D(0)} = \frac{1}{\eta_{A0}} \left(\frac{c}{d}\right)^2 \{\exp(\sigma^2) - 1\} , \tag{4.44}$$

while the half power width of the gaussian shaped error pattern is given by

$$\theta_E = 2\sqrt{\ln 2} \left(\frac{2\lambda}{\pi c}\right) = 1.06 \frac{\lambda}{c} . \tag{4.45}$$

The ratio of the power scattered into the error beam to that in the main beam can be found from these equations. The beam solid angle of a gaussian pattern profile is given by the simple relation (see Ch. 5.3.2)

$\Omega_b = 1.133\ \Theta_A{}^2$, where Θ_A is the half power beam width.

Approximating the main beam by a gaussian too, we obtain for the *ratio of the power in error to main beam*, using Eqs. (4.44), (4.45) and $\Theta_A = b\,\lambda\,/\,d$ with b = 1.2 (a typical value for practical tapers, see Fig. 4.6),

$$\frac{P_E}{P_M} = \frac{g_E(0)\,\Omega_E}{g_D(0)\,\Omega_M} \approx 1.1\,\{\exp(\sigma^2) - 1\}\,. \tag{4.46}$$

Note that this quantity is not dependent on the correlation length; it is solely determined by the rms surface error. From Eqs. (4.44) and (4.45) it is clear that a smaller correlation length will lead to a wider error pattern with a smaller peak level. For the observation of relatively small, be it extended, sources this will normally be preferable. However, even the very low error beam level, extending over a large solid angle can hamper observations of widely distributed radiation, as for instance the Galactic neutral hydrogen (Hartmann et al., 1996).

4.6.2. Numerical results with Mathematica

We begin with Eq. (4.40a) for the gain loss due to random errors, plotting the relative aperture efficiency, also called the surface error efficiency η_e, as function of wavelength for several error values (Fig. 4.21) [Mat.4.23]. We use d/c = 10 ($c^* = 0.1$) for this example; further we assume that η_{A0} = 0.72.This value is determined in the first place by the illumination efficiency (see Sec. 4.2), aperture blocking (see Sec. 4.5) and spillover (next Chapter). Values between 0.70 and 0.75 are obtained in practice for standard Cassegrain systems. In so-called shaped dual-reflector systems higher values can be achieved.

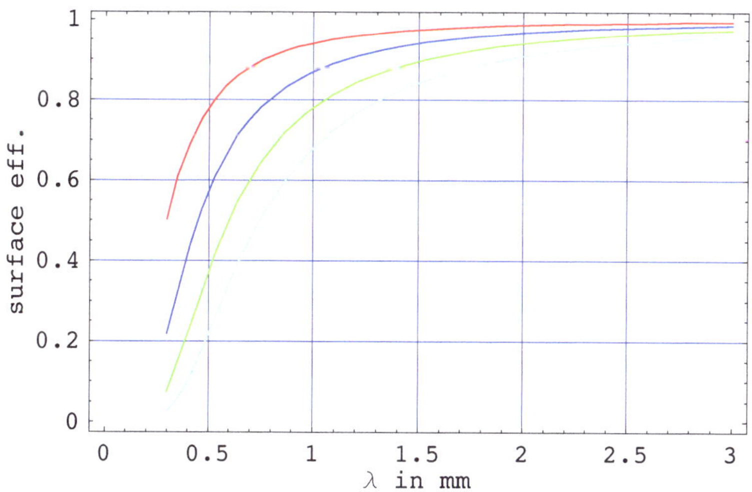

Fig. 4.21. Surface error efficiency η_e as function of wavelength (λ in mm) for four values of the rms error ϵ= 20, 30, 40 and 50 μm from top to bottom graph. Here d/c=10; larger values of this ratio make essentially no difference.

The following plot (Fig. 4.22) [Mat.4.24] shows the same relation, but now as function of the rms error with the wavelength as parameter. For simplicity, we have dropped the insignificant second term in Eq. (4.40) from the computation. Clearly these curves can be used for other wavelength ranges as long as the ratio of error to wavelength is kept constant. Thus in Fig. 4.22, if we use wavelength in cm in stead of mm, the error scale must be multiplied by a factor 10. A convenient single curve of the surface error efficiency as function of the ratio $\delta = \epsilon/\lambda$ is shown in Fig. 4.23 [Mat.4.25].

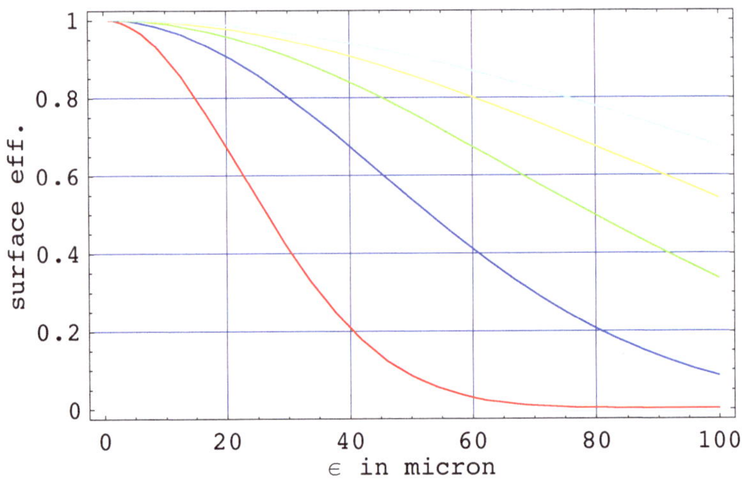

Fig. 4.22. Surface error efficiency η_e as function of the rms error ϵ (in μm) for 5 values of the wavelength, λ= 0.4, 0.8, 1.2, 1.6 and 2.0 mm from bottom to top.

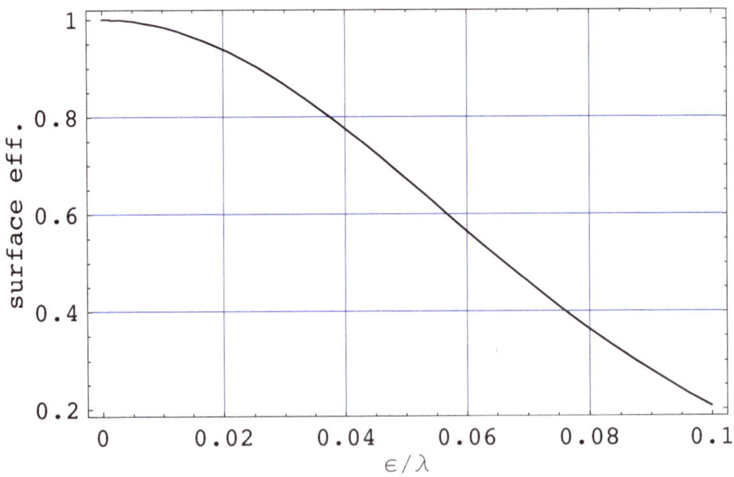

Fig. 4.23. Surface efficiency η_e as function of the ratio rms error to wavelength.

Now follows a set of curves, representing Eq. (4.44) [Mat.4.26], of the level of the error pattern $f_E(0)$ with respect to that of the (weakened) main beam $f_D(0)$ as function of ϵ/λ and for a number of values of the ratio $c^* = c/d$. From Fig. 4.24 one sees that for an rms error of less than $\lambda/20$ and a correlation length smaller than d/6 the error pattern level stays below approximately -20 dB, i.e. comparable with the normal sidelobe level of the main diffraction pattern. In conclusion, the relation of Eq. (4.46) [Mat.4.27], showing the ratio of the power in the error pattern to that in the main beam is depicted in Fig. 4.25. When the rms surface error approaches one twentieth of a wavelength, about thirty percent of the received power is distributed to the error pattern.

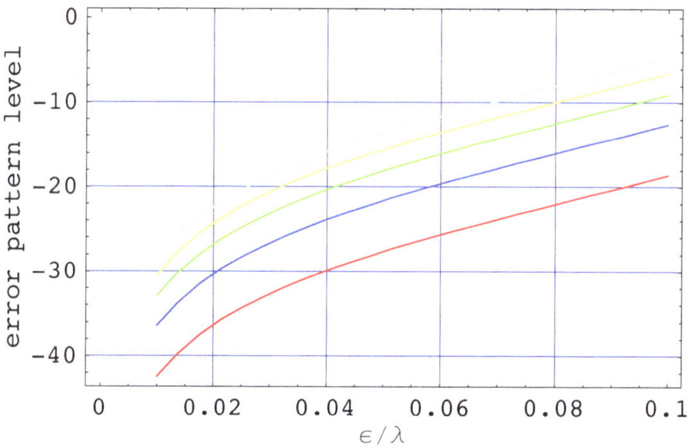

Fig. 4.24. The level of the error beam (in dB) with respect to the main beam as function of the ratio rms error to wavelength (ϵ/λ); parameter is the ratio c/d = 0.05 to 0.25 in steps of 0.05 from bottom.

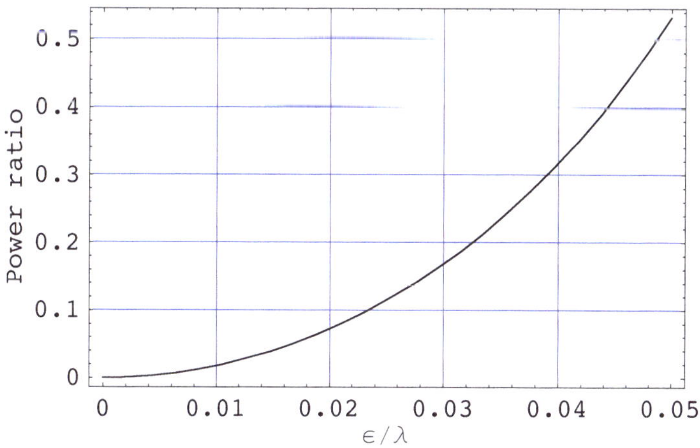

Fig. 4.25. The ratio of the power in the error pattern to that in the main beam as function of the surface error in units of the wavelength.

4.6.3. Large scale deformations - Astigmatism

The random errors, discussed in the foregoing section, represent mainly fabrication errors in the panels, constituting the reflector, and the errors in setting these panels onto the support structure with respect to the required surface contour. In addition to these errors with relatively small scale length, the reflector can exhibit large scale errors of a more systematic nature caused by, for instance, the varying gravitational forces on the structure as function of the elevation angle or by the asymmetric heating from solar radiation. An often observed deformation due to gravity is a so-called *astigmatism*, which is characterised by the "upper" and "lower" quadrant of the reflector having a focal length different from the "left" and "right" quadrants.

With changing elevation angle the amount of astigmatism is likely to vary because of the different orientation of the structural members of the antenna with respect to the gravity vector. As mentioned earlier (Eq. 2.22), astigmatism is characterised by a path length variation over the aperture of the form

$$\Delta_a = -\frac{\delta^2}{2f^3} r^2 \cos^2\chi = -\frac{\delta^2}{4f^3} r^2 (\cos 2\chi + 1) , \tag{4.47}$$

where χ is the azimuthal aperture coordinate. This additional phase term must be introduced in the radiation integral of Eq. (3.29). Assuming uniform illumination for simplicity, the radiation function now becomes

$$f(\theta,\ \phi) = \int_0^1 \int_0^{2\pi} \exp\left[ik\left\{-r\sin\theta\cos(\phi-\chi) - \frac{\delta^2}{4f^3} r^2 (\cos 2\chi + 1)\right\}\right] r\, dr\, d\chi . \tag{4.48}$$

The solution of this integral can be achieved most conveniently with the aid of Zernike polynomials. We shall not go through the entire development, which can be found in Born and Wolf (1980, Ch. 9.4). There we find that in the case of astigmatism, the radiation function takes the form

$$f(\theta,\ \phi) = \frac{2J_1(u)}{u} - (i\alpha_{022})\left(2\cos 2\phi\, \frac{2J_3(u)}{u}\right) + \tag{4.49}$$
$$(i\alpha_{022})^2 \frac{1}{2u}\left\{\frac{J_1(u)}{3} - \frac{J_3(u)}{2} + J_5(u)\left(\cos 4\phi + \frac{1}{6}\right)\right\} + \ldots .$$

Here α_{022} is a measure for the amount of astigmatism, related to the Seidel coefficients of aberration. The first term is the usual Lambda function for the uniformly illuminated aperture, the other terms describe the astigmatism. A result of the aberration is that the radiation pattern is no longer circularly symmetric. In particular, the focal point of the "left-right" quadrants will be positioned differently from that of the "upper-lower" quadrants. The optimum gain is at a point between these two partial foci, but the beam is distorted in that is exhibits different half-power beam widths in the two principal planes.

We illustrate this in Fig. 4.26 with the *Mathematica* expression [Mat.4.29], where the square of Eq. (4.49) is shown, i.e. the power pattern of the astigmatic aperture distribution. Clearly the sideobes are much stronger in the planes where the astigmatism is most relevant (red and blue curves). For the azimuthal angel $\phi=\pi/4$ the astigmatism is minimal and the beam is undistorted. The beamwidth in the different planes are slightly different, as shown in the lower part of the figure, where the level near the half-power point is shown.

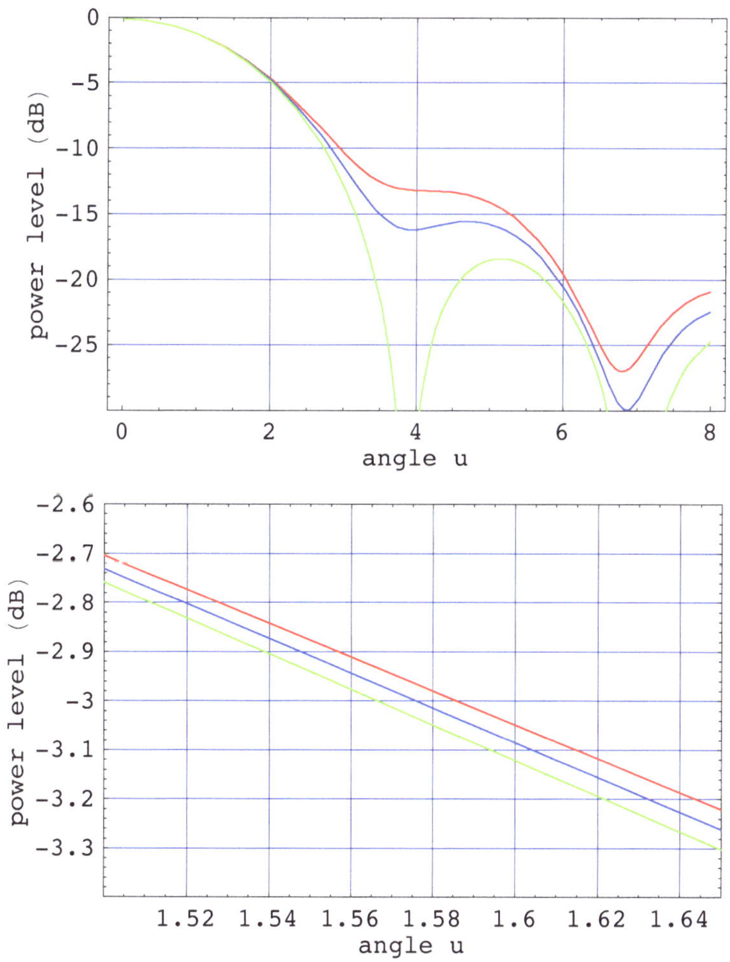

Fig. 4.26. Beam patterns with astigmatism for three values of the azimuthal angle ϕ. For $\phi=\pi/4$, green curve, the astigmatism is essentially zero. The other curves show significant sidelobe increase and some beam broadening. The latter is visible in the lower plot, exploded near the half-power point.

A measurement of the beam width at a number of azimuthal angles provides the information needed to evaluate the amount of astigmatism. A particularly powerful method is to use a feed with a "fan" beam, which illuminates only a strip of the reflector, and rotate the illumination pattern over the dish, while measuring the beam width. This was done successfully by von Hoerner (1978) on the NRAO 140-foot telescope. Without such a special device, good results can be obtained by measuring the beam width in two orthogonal planes for a number of axially defocused feed positions (Cogdell and Davis, 1973, Greve et al. 1994).

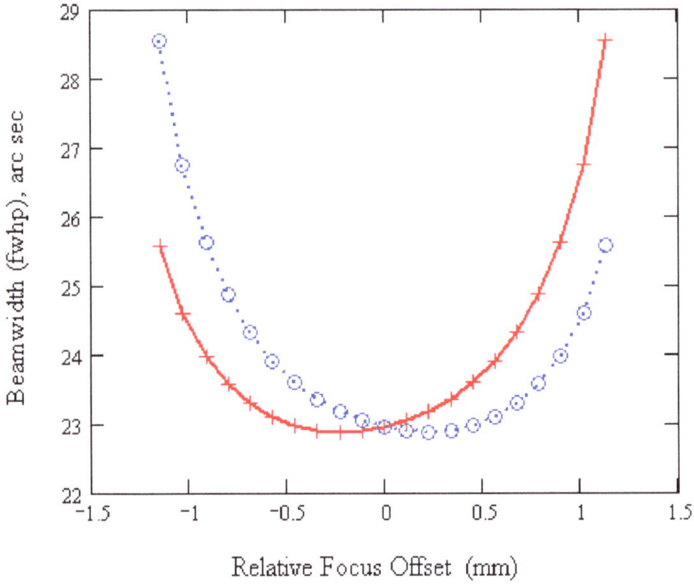

Fig. 4.27. Computed increase of beam width in the orthogonal azimuth (red) and elevation (blue) planes as function of axial defocus for a reflector with significant astigmatism. The minimum beam width occurs for different defocus positions in both planes (Emerson, 2006).

In a study of astigmatism Emerson (2006) found some interesting relations which we summarise here. Starting point is a model of a prime focus reflector with astigmatism as given in Eq. (4.49), to which the phase function of an axial defocus is added. Beam patterns are calculated for different axial defocus values and the beam width determined. Fig. 4.27 shows the computed beamwidth in the two orthogonal (azimuth and elevation) directions as function of axial defocus for a certain amount of astigmatism. As expected, the minimum beam width is found for different values of the focus offset.

Emerson then plotted the ratio of the beam widths in orthogonal planes as function of the axial defocus and found an approximate linear relationship. The **slope** of this line results effectively from a double differentiation, which will minimise the influence of non-astigmatic deformations. The beam width ratio as function of the axial defocus was then recalculated for a range of astigmatism values and the slopes of the resulting curves determined. Emerson then found that the slope of the curve representing the slopes of the beam width ratios is closely linearly proportional to the amplitude of the astigmatism.

In a practical situation we measure the beam width ratio as function of axial defocus and determine the slope of the curve. We can now compare this slope with the calculated "slope versus astigmatism" model function to find the astigmatism of the reflector. An example of such a measurement on an ALMA prototype antenna is shown in Fig. 4.28 (Emerson, 2006). The best fit to the measured points indicates an astigmatism of approximately 16 μm. This is a simple, yet powerful method to investigate astigmatism.

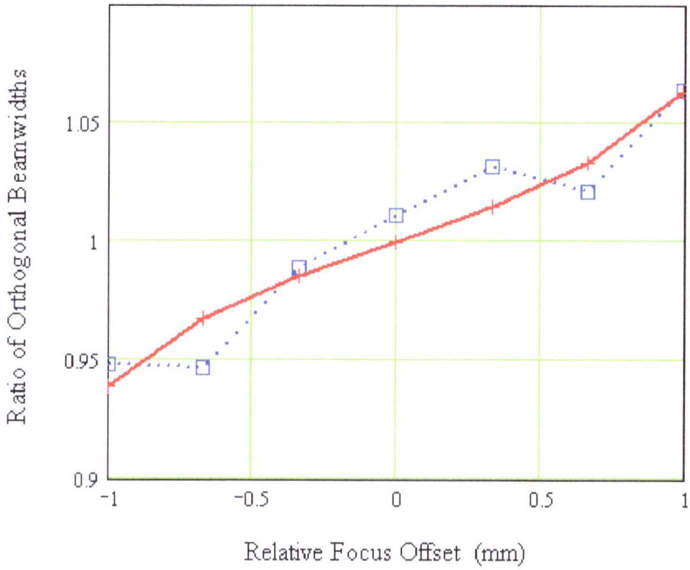

Fig. 4.28. Measured ratio of the beam widths in orthogonal planes as function of axial defocus for a reflector with astigmatism, (Blue squares) The best fit expected function (red) indicates an astigmatism of 16 μm.

In the following chapter we shall apply the results of our calculations to practical matters of determining the major parameters of an antenna from measurements using the radiation of cosmic radio sources. There particular attention will be paid to the situation where the test source has a finite angular size, comparable to that of the antenna. Because this is an often occurring situation in astronomical observations it is of special importance that the influence of the finite size be understood and corrected as well as possible.

■ 4.7. The *Mathematica* Routines

```
Mat .4 .1 - free space taper vs f / d - ratio;
t = 20 Log[10, (1 + (1 / (4 fr)) ^2)] ;
Plot[t, {fr, 0.25, 2}, Frame → True, GridLines → Automatic,
  FrameLabel → {"f/d", "free space taper (dB)"}]
```

```
Mat .4 .2 - taper and decibel equivalent;
Plot[20 Log[10, τ], {τ, 0, 1}, GridLines -> Automatic,
 Frame → True, PlotRange → {-35, 0},
 FrameLabel → {"edge taper", "taper in dB"}]
```

```
Mat .4 .3 - illumination functions;
ig = Exp[-1.3816 r^2];
ip = 1 - (1 - 0.251) r^2;
Plot[{ig, ip}, {r, 0, 1.2},
 Frame -> True, GridLines -> Automatic,
 FrameLabel -> {Radius, Illum. Function},
 PlotStyle -> {{RGBColor[1, 0, 0]}, {RGBColor[0, 0, 1]}}]
```

```
Mat .4 .4 - illumination efficiency;
jg = (2 (1 - Exp[(T Log[10] / 20)])^2) /
    ((-T Log[10] / 20) (1 - Exp[(T Log[10] / 10)]));
jp = (3 (1 + 10^(T/20))^2) / (4 (1 + 10^(T/20) + 10^(T/10)));
Plot[{jg, jp}, {T, 0, -30},
 Frame -> True, GridLines -> Automatic,
 PlotStyle -> {{RGBColor[1, 0, 0]}, {RGBColor[0, 0, 1]}},
 FrameLabel -> {"Taper in dB", "Illum. Efficiency"}]
```

```
Mat .4 .5 - farfield pattern parameters;
τ =.;
g = (4/(1 + τ) (τ BesselJ[1, u] / u + 2 (1 - τ) BesselJ[2, u] / u^2))^2;
Table[FindRoot[g == .5, {u, 1.5}], {τ, 1, 0, -.2}]
Table[FindMaximum[10 Log[10, g], {u, 5.19}], {τ, 1, 0, -.2}]
Plot[Evaluate[Table[10 Log[10, g], {τ, 1, 0, -.2}]],
 {u, 0, 10}, Frame -> True, GridLines -> Automatic,
 FrameLabel -> {"u", "Gain (dB)"}, PlotRange → {-30, 0},
 PlotStyle → {{RGBColor[1, 0, 0]}, {RGBColor[0, 1, 0]},
   {RGBColor[0, 0, 1]}, {RGBColor[1, 1, 0]},
   {RGBColor[1, 0, 1]}, {RGBColor[0, 1, 1]}}]
```

```
Mat .4 .6 - beamwidth factor;
da =
   {1.61634, 1.64895, 1.69239, 1.75311, 1.84397, 1.99442};
dat = (2 / π) da;
data = {{1, dat[[1]]}, {.8, dat[[2]]}, {.6, dat[[3]]},
    {.4, dat[[4]]}, {.2, dat[[5]]}, {0, dat[[6]]}};
fsl3 = Fit[data, {1, x, x^2, x^3}, x]
psl = ListPlot[data, PlotStyle → PointSize[0.015]]
pfsl = Plot[fsl3, {x, 0, 1}, DisplayFunction → Identity]
Show[{psl, pfsl}, PlotRange → {1, 1.3}, Frame → True,
 GridLines → Automatic, FrameLabel → {"taper", "factor b"}]
```

Out[82]= $1.26914 - 0.566036\, x + 0.53346\, x^2 - 0.208066\, x^3$

```
Mat .4 .7 - sidelobe level vs taper;
data = {{0, -24.6392}, {.2, -23.4225}, {.4, -21.4776},
    {.6, -19.8295}, {.8, -18.555}, {1, -17.57}};
fsl3 = Fit[data, {1, x, x^2, x^3}, x]
fsl1 = Fit[data, {1, x}, x]
psl = ListPlot[data, PlotStyle → PointSize[0.015]]
pfsl = Plot[{fsl1, fsl3}, {x, 0, 1},
   PlotStyle → {{RGBColor[1, 0, 0]}, {RGBColor[0, 0, 0]}},
   DisplayFunction → Identity]
Show[{psl, pfsl}, PlotRange → {-25, -17},
 Frame → True, GridLines → Automatic,
 FrameLabel → {"taper", "sidelobe (dB)"}]
```

Out[88]= $-24.6819 + 5.71171\, x + 7.51987\, x^2 - 6.15613\, x^3$

Out[89]= $-24.6011 + 7.37094\, x$

```
Mat .4 .8 - axial defocus gain function; τ =.;
fc = Integrate[(2 / (1 + τ)) (1 - (1 - τ) z) Cos[b z], {z, 0, 1}];
fs = Integrate[(2 / (1 + τ)) (1 - (1 - τ) z) Sin[b z], {z, 0, 1}];
ga = fc^2 + fs^2 ;
TrigExpand[ga];
gas = FullSimplify[%]
```

Out[98]= $\frac{1}{b^4\,(1+\tau)^2}\,(4\,(2+b^2-4\,\tau+(2+b^2)\,\tau^2-2\,(1+\tau\,(-2+b^2+\tau))\,\mathrm{Cos}[b]-2\,b\,(-1+\tau)^2\,\mathrm{Sin}[b]))$

```
Mat .4 .9 - axial defocus approximations;
δ =.; f = 0.8; λ = 0.003; k = 2 π / λ;
glc = NIntegrate[
    2 Cos[k (Sqrt[r^2 + (f - r^2/(4 f) + δ)^2] - (f + r^2/(4 f) + δ))] r,
    {r, 0, 1}];
gls = NIntegrate[
    2 Sin[k (Sqrt[r^2 + (f - r^2/(4 f) + δ)^2] - (f + r^2/(4 f) + δ))] r,
    {r, 0, 1}];
gbc = NIntegrate[2 Cos[k δ 2 (r/(2 f))^2 / (1 + (r/(2 f))^2)] r,
    {r, 0, 1}];
gbs = NIntegrate[2 Sin[k δ 2 (r/(2 f))^2 / (1 + (r/(2 f))^2)] r,
    {r, 0, 1}];
gl = (glc^2 + gls^2);
gb = (gbc^2 + gbs^2);
gs = (Sin[1.7649 δ / λ] / (1.7649 δ / λ)) ^2;
gp = 1 - ((k δ 0.5618) ^2) / 12;
Plot[{gp, gl, gb, gs}, {δ, 0, .015},
 PlotRange → {0, 1}, Frame -> True, GridLines -> Automatic,
 FrameLabel -> {"Axial defocus (m)", "Relative Gain"},
 PlotStyle → {{RGBColor[0, 1, 1]}, {RGBColor[1, 0, 0]},
   {Dashing[{.05, 0.05}], RGBColor[0, 1, 0]},
   {RGBColor[0, 0, 1]}}]
```

```
Mat .4 .10 - axial defocus gain plots;
                1
gax = ─────────────── (4 (2 + b^2 - 4 τ + (2 + b^2) τ^2 -
        b^4 (1 + τ)^2
          2 (1 + τ (-2 + b^2 + τ)) Cos[b] - 2 b (-1 + τ)^2 Sin[b]));
Plot[Evaluate[Table[gax, {τ, 0, 1, .25}], {b, 0, 20},
   PlotRange → {0, 1}, GridLines → Automatic,
   Frame → True, FrameLabel → {"defocus-β", "gain"},
   PlotStyle → {{RGBColor[1, 0, 0]},
      {RGBColor[0, 1, 0]}, {RGBColor[0, 1, 1]},
      {RGBColor[1, 0, 1]}, {RGBColor[0, 0, 1]}}]]
Plot[Evaluate[Table[10 Log[10, gax], {τ, 0, 1, .25}],
   {b, 0, 20}, PlotRange → {-30, 0}, GridLines → Automatic,
   Frame → True, FrameLabel → {"defocus-β", "gain (dB)"},
   PlotStyle → {{RGBColor[1, 0, 0]},
      {RGBColor[0, 1, 0]}, {RGBColor[0, 1, 1]},
      {RGBColor[1, 0, 1]}, {RGBColor[0, 0, 1]}}]]
```

```
Mat .4 .11 - axial defocus beam patterns;
δ =.; λ = 0.003; k = 2 π / λ; fr = 1;
fac =
   Integrate[ BesselJ[0, u r] Cos[0.5618 k δ r^2] r , {r, 0, 1}];
fas = Integrate[BesselJ[0, u r] Sin[0.5618 k δ r^2] r, {r, 0, 1}];
ga = 4 (fac^2 + fas^2);
Plot[Evaluate[Table[10 Log[10, ga], {δ, 0, .005, .001}],
   {u, 0, 10}, PlotRange -> {-40, 0},
   Frame -> True, GridLines -> Automatic,
   FrameLabel -> {"angle u", "gain (dB)"},
   PlotStyle -> {{RGBColor[1, 0, 0]}, {RGBColor[0, 1, 0]},
      {RGBColor[0, 0, 1]}, {RGBColor[1, 1, 0]},
      {RGBColor[1, 0, 1]}, {RGBColor[0, 1, 1]}}]]
```

```
Mat .4 .12 - axial defocus beam patterns;
δ =.; λ = 0.003; k = 2 π / λ; fr = 1 - 0.75 r^2;
fac = Integrate[
    1.6 fr BesselJ[0, u r] Cos[0.5618 k δ r²] r , {r, 0, 1}];
fas = Integrate[1.6 fr BesselJ[0, u r] Sin[0.5618 k δ r²] r,
    {r, 0, 1}];
ga = 4 (fac^2 + fas^2);
Plot[Evaluate[Table[10 Log[10, ga], {δ, 0, .005, .001}],
   {u, 0, 10}, PlotRange -> {-40, 0},
   Frame -> True, GridLines -> Automatic,
   FrameLabel -> {"angle u", "gain (dB)"},
   PlotStyle -> {{RGBColor[1, 0, 0]}, {RGBColor[0, 1, 0]},
     {RGBColor[0, 0, 1]}, {RGBColor[1, 1, 0]},
     {RGBColor[1, 0, 1]}, {RGBColor[0, 1, 1]}}]]
```

```
Mat .4 .13 - lateral defocus beam patterns;
Ξ = 1.6; d = 12; λ = 0.003; k = 2 π / λ;
f10 = NIntegrate[
    2 r * BesselJ[0, r (u - 2 k 0 / (Ξ (1 + (r/Ξ)^2)))] , {r, 0, 1}];
f11 = NIntegrate[2 r * BesselJ[0, r (u - 2 k .001 / (Ξ (1 + (r/Ξ)^2)))] ,
    {r, 0, 1}];
f12 = NIntegrate[2 r * BesselJ[0, r (u - 2 k .002 / (Ξ (1 + (r/Ξ)^2)))] ,
    {r, 0, 1}];
f13 = NIntegrate[2 r * BesselJ[0, r (u - 2 k .003 / (Ξ (1 + (r/Ξ)^2)))] ,
    {r, 0, 1}];
fl = {f10^2, f11^2, f12^2, f13^2};
Plot[Evaluate[10 Log[10, fl], {u, -10, 20},
   PlotRange -> {-30, 0}, Frame -> True, GridLines -> Automatic,
   PlotStyle -> {{RGBColor[1, 0, 0]}, {RGBColor[0, 1, 0]},
     {RGBColor[0, 0, 1]}, {RGBColor[1, 0, 1]}},
   FrameLabel -> {"angle u", "Gain (dB)"}]]
```

```
Mat .4 .14 - off - axis beam plot;
λ = 0.003; k = 2 π / λ; Ξ = 1.6;
i1 = NIntegrate[
    2 r * BesselJ[0, ((x - (2 k 0.003)/(Ξ (1 + (r/Ξ)^2)))^2 + y^2)^.5 r], {r, 0, 1}];
ContourPlot[Evaluate[i1^2, {x, -5, 15}, {y, -10, 10},
   ContourLines -> False, DisplayFunction → Identity]]
Plot3D[i1^2, {x, -5, 20}, {y, -10, 10}, BoxRatios → {1, 1, 1},
 PlotRange → All, DisplayFunction → Identity]
Show[GraphicsArray[{%, %%}]]
```

```
Mat .4 .15 - lateral defocus beam patterns;
Ξ = 1.6; d = 12; λ = 0.003; k = 2 π / λ; fr = 2 r (1 - .75 r^2);
f10 = NIntegrate[
    fr * BesselJ[0, r (u - (2 k 0)/(Ξ (1 + (r/Ξ)^2)))] , {r, 0, 1}];
f11 = NIntegrate[fr * BesselJ[0, r (u - (2 k 0.001)/(Ξ (1 + (r/Ξ)^2)))] ,
    {r, 0, 1}];
f12 = NIntegrate[fr * BesselJ[0, r (u - (2 k 0.002)/(Ξ (1 + (r/Ξ)^2)))] ,
    {r, 0, 1}];
f13 = NIntegrate[fr * BesselJ[0, r (u - (2 k 0.003)/(Ξ (1 + (r/Ξ)^2)))] ,
    {r, 0, 1}];
FindMaximum[10 Log[10, 2.56 f10^2], {u, -5}]
FindMaximum[2.56 f10^2, {u, 0}]
similar expressions for the other beams;
fl = {f10^2, f11^2, f12^2, f13^2};
Plot[Evaluate[10 Log[10, 2.56 fl], {u, -10, 15},
  PlotRange -> {-30, 0}, Frame -> True, GridLines -> Automatic,
  PlotStyle -> {{RGBColor[1, 0, 0]}, {RGBColor[0, 1, 0]},
     {RGBColor[0, 0, 1]}, {RGBColor[1, 0, 1]}},
  FrameLabel -> {"angle u", "Gain (dB)"}]]
```

```
Mat .4 .16 - Beam Deviation Factor;
Plot[
  Evaluate[Table[Integrate[((1 - (1 - τ) r^2) r^3)/(1 + (r/(4 f))^2), {r, 0, 1}] /
      Integrate[(1 - (1 - τ) r^2) r^3, {r, 0, 1}], {τ, 0, 1, .25}],
   {f, 0.25, 1}, Frame → True, GridLines → Automatic,
   FrameLabel → {focal ratio, BeamDeviationFactor},
   PlotStyle → {{RGBColor[1, 0, 0]},
     {RGBColor[0, 0, 1]}, {RGBColor[1, 1, 1]},
     {RGBColor[1, 1, 1]}, {RGBColor[0, 1, 0]}}]]
```

```
Mat .4 .17 - BDF vs taper;
n1 = ∫_0^1 ((1 - (1 - 10^(-t/20)) r^2) r^3)/(1 + (r/(4 f))^2) dr;
d1 = ∫_0^1 (1 - (1 - 10^(-t/20)) r^2) r^3 dr;
BDF = n1 / d1;
Plot[Evaluate[Table[BDF, {f, .35, .5, .05}],
   {t, 0, 30}, Frame -> True, GridLines → Automatic,
   FrameLabel → {"taper (dB)", "BDF"},
   PlotStyle -> {{RGBColor[1, 0, 0]}, {RGBColor[0, 1, 0]},
     {RGBColor[0, 0, 1]}, {RGBColor[1, 0, 1]}}]]
```

```
Mat .4 .18 - BDF vs f / d and taper;
Plot3D[Evaluate[BDF, {f, .25, 1}, {t, 0, 30},
   ViewPoint → {-2, -2, 0}, BoxRatios → {1, 1, .75}, AxesLabel →
    {"f/D-ratio", "Taper(dB)", "BDF"}, Mesh → True]]
```

Mat.4.19 - Blocking input parameters;

$Rp = 6.0;\ Rs = 0.375;\ Rq = 4.11;\ F = 4.8;$

$Wl = 0.06;\ n = 4;\ \alpha = 42.89\,\pi / 180;$

$\Psi_0 = 2\,\mathrm{ArcTan}[Rp / (2\ F)]$

$\psi = 2\,\mathrm{ArcTan}[Rq / (2\ F)]$

$\tau = 0.75;$ (this gives 12 dB edge taper);

Mat.4.20 - blocking formulae;

$$Ar = \pi\, Rp^2$$

$$As = \pi\, Rs^2$$

$$Abp_u = n\, Wl\ (Rq - Rs)$$

$$Abp_t = \frac{n\, Wl\ (1 - \tau)}{3\ Rp^2}\ (Rq^3 - Rs^3)$$

$$AB = Rq\ (1 - \mathrm{Tan}[\alpha] / \mathrm{Tan}[\psi])$$

$$Abs_u = \frac{n\, Wl}{AB}\left(\frac{Rp^2 - Rq^2}{2} - (Rp - Rq)\ F\,\mathrm{Tan}[\alpha] + \frac{\mathrm{Tan}[\alpha]}{12\ F}\ (Rp^3 - Rq^3)\right)$$

$$Abs_t = \frac{n\, Wl\ (1 - \tau)}{AB\ Rp^2}\left(\frac{(Rp^4 - Rq^4)}{4} - F\,\mathrm{Tan}[\alpha]\ \frac{(Rp^3 - Rq^3)}{3} + \frac{\mathrm{Tan}[\alpha]}{20\ F}\ (Rp^5 - Rq^5)\right)$$

$$Abs_a = (n\, Wl / 2)\ (Rp - Rq)\left(1 + \frac{Rp}{\mathrm{Sin}[\Psi_0]} \Big/ \left(\frac{\mathrm{Sin}[\alpha]}{\mathrm{Sin}[\Psi_0 - \alpha]}\left(\frac{Rq^2}{4\ F} + \frac{Rq}{\mathrm{Tan}[\alpha]} - F\right)\right)\right)$$

Mat.4.21 - blocking final calculation;

$Ab = As + Abp_u + Abs_u$

$BlPer = 100\ \% / Ar$

$Ab_a = As + Abp_u + Abs_a$

$BlPer = 100\ \% / Ar$

$Ab_t = Ab - Abp_t - Abs_t$

$BlPer_t = 100\ \% / Ar$

$Ab_{ta} = Ab_a - Abp_t - Abs_t$

$BlPer_{at} = 100\ \% / Ar$

```
Mat .4 .22 - Beam of annular apertures;
fu = 2 BesselJ[1, u] / u;
fb = Integrate[2 BesselJ[0, u r] r, {r, 0.2, 1}];
fl = Integrate[2 BesselJ[0, u r] r, {r, 0.95, 1}];
Plot[Evaluate[{fu^2, fb^2, 100 fl^2}],
  {u, 0, 15}, PlotRange → All, Frame → True,
  FrameLabel → {"angle u", "Power (dB)"},
  PlotStyle -> {{RGBColor[1, 0, 0]},
    {RGBColor[0, 0, 1]}, {RGBColor[0, 1, 0]}}]
```

```
Mat .4 .23 - full "Ruze" formula;
λ =.; c* = 0.1; ηAO = 0.72;
Plot[Evaluate[
   Table[(Exp[-(4 π ϵ/(100 * λ))^2] + 1/ηAO (c*)^2 (1 - Exp[-(4 π ϵ/(100 * λ))^2])),
    {ϵ, 2, 5, 1}]], {λ, .3, 3},
  FrameLabel -> {"λ in mm", "surface eff."},
  Frame -> True, GridLines -> Automatic,
  PlotStyle → {{RGBColor[1, 0, 0]}, {RGBColor[0, 0, 1]},
    {RGBColor[0, 1, 0]}, {RGBColor[0, 1, 1]}}]
```

```
Mat .4 .24 - "Ruze" efficiency vs error;
Plot[Evaluate[Table[Exp[-(4 * π * ϵ/(1000 * λ))^2], {λ, .4, 2, .4}]],
  {ϵ, 0, 100}, FrameLabel -> {"ϵ in micron", "surface eff."},
  Frame -> True, GridLines -> Automatic,
  PlotStyle → {{RGBColor[1, 0, 0]},
    {RGBColor[0, 0, 1]}, {RGBColor[0, 1, 0]},
    {RGBColor[1, 1, 0]}, {RGBColor[0, 1, 1]}}]
```

```
Mat .4 .25 - the "Ruze" surface error formula;
Plot[Exp[-(4 π δ)^2], {δ, 0, 0.1}, Frame → True,
  GridLines → Automatic, FrameLabel → {"ϵ/λ", "surface eff."}]
```

```
Mat .4 .26 - error pattern level;
Plot[Evaluate[Table[10 Log[10, 1.4 * c*^2 * (Exp[(4 π δ)^2] - 1)],
    {c*, 0.05, 0.25, 0.05}]], {δ, .01, 0.1},
  Frame -> True, GridLines -> Automatic,
  FrameLabel -> {"ε/λ", "error pattern level"},
  PlotStyle -> {{RGBColor[1, 0, 0]},
    {RGBColor[0, 0, 1]}, {RGBColor[0, 1, 0]},
    {RGBColor[1, 1, 0]}, {RGBColor[0, 1, 1]}}]
```

```
Mat .4 .27 - relative power in error pattern;
Plot[1.1 (Exp[(4 π δ)^2] - 1), {δ, 0, .05},
  GridLines -> Automatic, Frame -> True,
  FrameLabel -> {"ε/λ", "Power ratio"}]
```

```
Mat .4 .28 - 3 D plots of efficiency and error pattern;
p1 =
  Plot3D[Evaluate[Exp[-(4 π δ)^2] + 1.4 (c*)^2 (1 - Exp[-(4 π δ)^2])],
    {δ, 0, .1}, {c*, .025, .3}, ViewPoint → {2, -2, 0},
    BoxRatios → {1, 1, .75}, DisplayFunction → Identity,
    AxesLabel → {"ε/λ", "c/d", "η"}]]
p2 = Plot3D[Evaluate[10 Log[10, 1.4 * c*^2 * (Exp[(4 π δ )^2] - 1)],
    {c*, 0.025, 0.3}, {δ, .01, 0.1}, ViewPoint → {-2, -2, 1},
    BoxRatios → {1, 1, .8}, DisplayFunction → Identity,
    AxesLabel → {"c/d", "ε/λ", "f_E"}]]
Show[GraphicsArray[{p1, p2}]]
```

```
Mat .4 .29 - beam with astigmatism;
α = .5; ϕ =.;
Plot[
  Evaluate[Table[10 Log[10, (2 BesselJ[1, u] / u - (α^2 / (2 u))
              (BesselJ[1, u] / 3 - BesselJ[3, u] / 2 +
                BesselJ[5, u] (Cos[4 ϕ] + 1 / 6))) ^2 +
          (α 4 Cos[2 ϕ] BesselJ[3, u] / u) ^2],
     {ϕ, 0, 2 π / 8, π / 8}], {u, 0, 8}, PlotRange →
     {-30, 0},
   Frame -> True, GridLines -> Automatic,
   FrameLabel →
     {"angle u", "power level (dB)"},
   PlotStyle → {{RGBColor[1, 0, 0]},
      {RGBColor[0, 0, 1]}, {RGBColor[0, 1, 0]}}]]
Show[{%}, PlotRange -> {{1.5, 1.65}, {-3.4, -2.6}},
  FrameLabel → {"angle u", "power level (dB)"}]
```

References

Baars, J.W.M., *Dual-beam parabolic antennae in radio astronomy*, Groningen, Wolters-Noordhof, 1970.

Baars, J.W.M., The Measurement of large Antennas with cosmic Radio Sources, *IEEE Trans. Antennas Propagat.* **AP-21**, 461-474, 1973 [Gold, 171].

Born, M. and E. Wolf, *Principles of Optics*, 6th Ed. Oxford, Pergamon, 1980.

Bracewell, R.N., Tolerance theory of large antennas, *IEEE Trans. Antennas Propagat.* **AP-9**, 49-58, 1961.

Cheng, J. and J.G. Mangum, Feed leg blockage and ground radiation pickup for Cassegrain antennas, MMA - Memo 197, 1998.

Cogdell, J.R. and J.H. Davis, Astigmatism in reflector antennas, *IEEE Trans. Antennas Propagat.* **AP-21**, 565-567, 1973.

Emerson, D., Astigmatism measurements, NRAO/ALMA Internal Report, 2006

Greve, A. and B.G. Hooghoudt, Quality evaluation of radio reflector surfaces, *Astron.Astrophys.* **93**, 76-78, 1981 [Gold, 197].

Greve, A., Strehl number degradation by large-scale systematic surface deviations (on optical and radio reflectors), *Appl. Opt.* ***19***, 2948-2951, 1980.

Greve, A., B. LeFloch, D. Morris, H. Hein and S. Navarro, Astigmatism in reflector antennas - Measurement and correction, *IEEE Trans Antennas Propag.* **42**, 1345-1350, 1994.

Hartmann, D., P.M.W. Kalberla, W.B. Burton and U. Mebold, Stray radiation correction as applied to the Leiden/Dwingeloo survey of HI in the Galaxy, *Astron. Astrophys. Suppl. Ser.* **119**, 115-151, 1996.

Hoerner, S. von, Measuring the gravitational astigmatism of a radio telescope, *IEEE Trans. Antennas Propagat.* **AP-26**, 315-318, 1978.

Jahnke, E. and F. Emde, *Tables of functions*, 4th Ed., New York, Dover, 1945.

Kraus, J.D., *Radio Astronomy*, New York, McGraw-Hill, 1966.

Maanders, E.J., *Enige aspecten van grondstation Antennes voor Satelliet Communicatie*, Dissertation University of Gent, Belgium, 1975.

Rusch, W.V.T., O. Sorensen and J.W.M. Baars, Radiation cones from feed-support struts of symmetric paraboloidal antennas, *IEEE Trans. Antennas Propag.* **AP-30**, 786-790, 1982.

Ruze, J., The effect of aperture errors on the antenna radiation pattern, *Suppl. al Nouvo Cimento* **9**, 364-380, 1952.

Ruze, J., Lateral feed displacement in a paraboloid, *IEEE Trans. Antennas Propagat.* **AP-13**, 660-665, 1965.

Ruze, J., Antenna tolerance theory - a review, *Proc. IEEE* **54**, 633-640, 1966 [Gold, 185].

Ruze, J., Feed support blockage loss in parabolic antennas, *Microwave J.*, pp. 76-80, Dec. 1968 [Gold, 77].

Scheffler, H., Über die Genauigkeitsanforderungen bei der Herstellung optischer Flächen für astronomische Teleskope, *Z. Astrophys.* **55**, 1-20, 1962.

Silver, S., *Microwave Antenna Theory and Design*, MIT Rad. Lab Series **12**, New York, McGraw-Hill, 1949.

5. Measurement of antenna parameters

5.1. Global antenna parameters

In Chapter 3 we discussed the mathematical description of the radiation characteristics of the reflector antenna, and we treated the influence of aberrations (defocus situations) and other errors in Ch. 4. In radio astronomy the purpose of the telescope, the reflector antenna in our discussion, is to collect radiation from the celestial source as a function of position on the sky, frequency, polarisation and sometimes time. In order to draw conclusions about the source of radiation, we need to establish the relationship between the parameters describing the physical processes in the source and those of the receiving antenna. Thus we must develop a mathematical formulation for the interaction between the transmitting cosmic radio source (or satellite, etc) and the receiving radio telescope, the reflector antenna discussed sofar. As we have seen earlier, the characteristics of the latter are described by the spatial radiation characteristic, which we have called the *antenna pattern*. By virtue of the reciprocity theorem the pattern of a transmitting antenna is identical to that of a receiving antenna. (Silver, Ch. 2.13, 1949). We have already introduced terms like main beam and sidelobes of the antenna pattern.

We shall now define these parameters more carefully and develop a number of relations which are essential for the description of the interaction between antenna and source under study. Note that in all of the following discussion we are dealing with the **antenna power pattern**, because this is the quantity used in practice. The power pattern is often called the antenna **gain** function, and we denote it by $\mathbf{g}(\boldsymbol{\theta}, \boldsymbol{\phi})$, where θ and ϕ are the angular pattern coordinates as defined in Chapter 3. We also normalise the pattern to $g(0, 0) = 1$. The pattern will of course be dependent on the frequency. If not explicitly noted, we shall assume in the following discussion a situation of monochromatic radiation at frequency ν or radiation over a small bandwidth $\Delta\nu << \nu$. Although the majority of cosmic sources radiates over a wide, continuous bandwidth, the receiver system is normally only sensitive over a small bandwidth compared to the center reception frequency.

We commence with the definition and mutual relationships of a number of important antenna parameters.

The *directivity* of the antenna $D(\theta, \phi)$ is defined as the power received (or emitted) per unit solid angle in the direction (θ, ϕ) divided by the average power over a unit solid angle:

$$D(\theta, \phi) = \frac{4\pi\, g(\theta,\phi)}{\iint_{4\pi} g(\theta,\phi)\, d\Omega} . \tag{5.1}$$

Clearly, the maximum directivity will be in the direction of the main beam axis; this is often called "the directivity" D_M of the antenna. The integral in the denominator is called the *effective antenna solid angle* (or pattern solid angle), written as

$$\Omega_A = \iint_{4\pi} g(\theta, \phi)\, d\Omega . \tag{5.2}$$

Obviously we have
$$D_M = 4\pi / \Omega_A . \tag{5.3}$$

Another antenna parameter is also in use, called the *gain* $G(\theta, \phi)$. It is defined as

$$G(\theta, \phi) = \eta_R\, D(\theta, \phi), \tag{5.4}$$

where η_R is the *radiation efficiency* of the antenna, which represents the ohmic losses in the antenna reflector and feed. Normally η_R is very close to one and G and D are essentially identical. The directivity is a parameter which emerges from the consideration of the antenna as a transmitter. Alternatively, one can characterise the antenna from reception considerations and introduce the *effective absorption area $A(\theta, \phi)$*, defined as the power available at the antenna terminals divided by the power crossing a unit area of infalling wave front. It is, like the directivity and the beam pattern, a function of the angle from the beam axis. Normally, we use only the value A(0, 0) and call it the effective reflector area, denoted A.

Because of the reciprocity between transmitting and receiving parameters, D and A must be linearly related. We can show this by the following thermodynamic consideration. Consider an antenna with a matched resistor at temperature T connected to its terminals and place a power generator in series with the resistor, as illustrated in Fig. 5.1. From the reciprocity theorem it follows that there will be power transfer from any received wave to the resistor and from the power generator to the antenna, which is radiated. Now place a blackbody, also at temperature T, in space subtending a solid angle Ω at the antenna in a direction in which the gain is G. We can now write for the power radiated by the antenna towards the black body in the frequency interval $\Delta\nu$

$$P_t = k\,T\,\Delta\nu\, G\,\Omega / 4\pi, \tag{5.5a}$$

where $k = 1.38 \,.\, 10^{-23}\ W\,K^{-1}\,\mathrm{Hz}^{-1}$ is Boltzmann's constant. It should be borne in mind that this power is radiated in the polarisation state of the antenna. On the other hand, in the Rayleigh-Jeans part of the spectrum ($h\nu << kT$) the radiating blackbody has a brightness $B = 2\,k\,T\,/\,\lambda^2$ and emits randomly polarised radiation.

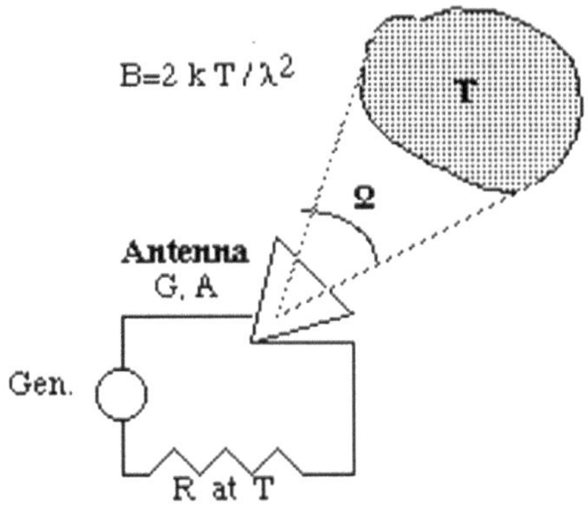

Fig. 5.1. Illustrating the reciprocity between reception and transmission of radiation by the antenna

The power received by the antenna in the solid angle Ω, absorbed by the effective area A and delivered to the matched load will be

$$P_r = \frac{1}{2}\,\frac{2\,k\,T\,\Delta\nu}{\lambda^2}\,\Omega\,A\,, \tag{5.5b}$$

where the factor one-half reflects the fact that only half the received power is matched to the polarisation state accepted by the antenna. Because all is at the same temperature, there is thermal equilibrium and the principle of detailed balancing requires that $P_t = P_r$. Thus we obtain

$$G = 4\pi\,A\,/\,\lambda^2\,. \tag{5.6}$$

It is interesting to note that we have not made any detailed specification of the type of antenna. This relation applies to any antenna and we can even use it to assign a value A to an antenna with known gain, even if the physical capture area is hard to discern, as e.g. in a dipole.

From the definitions above we can write the following useful relations between different quantities

$$\frac{\eta_R}{\Omega_A} = \frac{\eta_R\,D_M}{4\pi} = \frac{G}{4\pi} = \frac{A}{\lambda^2}\,. \tag{5.7}$$

■ 5.2. Response of an antenna to a source distributed in space

Consider an antenna receiving radiation from an extended source at large distance which has a brightness $B(\theta, \phi)$. The power received from a solid angle $d\Omega(\theta,\phi)$ by the aperture with absorption area $A(\theta, \phi)$ over a unit bandwidth is

$$\Delta p = B(\theta, \phi)\, A(\theta, \phi)\, d\Omega \quad (\mathrm{W\,Hz^{-1}})\,.$$

Integrating over the entire sky we obtain

$$p = \int_{4\pi} B(\theta, \phi)\, A(\theta, \phi)\, d\Omega. \tag{5.8}$$

The received power p can be expressed as the temperature T_A of a matched resistor at the antenna terminals ("radiation resistance") through the Nyquist (1928) formula

$$p = k\, T_A \quad (\mathrm{W\,Hz^{-1}}), \tag{5.9}$$

where $k = 1.38 \,.\, 10^{-23}\ W\,K^{-1}\ \mathrm{Hz}^{-1}$ is Boltzmann's constant.

We call this fictitious temperature the *antenna temperature*. Note that the antenna temperature is a measure of the received power by the antenna and is not related to the physical temperature of the antenna structure.

In a similar fashion we can express the source brightness in terms of a temperature, which we call *brightness temperature*, denoted by T_b. It is defined as the equivalent Rayleigh-Jeans blackbody temperature:

$$T_b = (\lambda^2 / 2\, k)\, B \ \ (\mathrm{K}) \tag{5.10}$$

The brightness temperature is also a fictitious quantity and will not be related to the physical temperature of the source apart from the case where the source can be characterised as a blackbody in the Rayleigh-Jeans approximation (e.g. the planets at radio wavelengths). Before we link the expressions for the power emitted by the source and that received by the antenna, we must remember that only half of the randomly polarised blackbody radiation will be received by the linearly (or circularly) polarised antenna. Combining Eqs. (5.8 - 5.10) we obtain

$$T_A = \lambda^{-2} \int_{4\pi} T_b(\theta, \phi)\, A(\theta, \phi)\, d\Omega. \tag{5.11}$$

An interesting result can be obtained by the following argument, which is essentially identical to that made above in Ch. 5.1. Let the antenna, terminated by a matched resistor, be surrounded completely by a surface at temperature T. In a state of equilibrium the antenna temperature will be $T_A = T$, while also $T_b = T$. Substitution in Eq. (5.11) leads to

$$\int_{4\pi} A(\theta, \phi)\, d\Omega = \lambda^2, \tag{5.12}$$

Thus the full sphere integral of the reception pattern is equal to the wavelength squared. We can write Eq. (5.12) also as

$$A(0, 0) = \lambda^2 \Big/ \int_{4\pi} g(\theta, \phi)\, d\Omega = \lambda^2 / \Omega_A = \lambda^2 G / 4\pi, \tag{5.13}$$

where we have made use of Eqs. (5.2 and 5.3). It is easily seen that Eq. (5.13) is identical to Eq. (5.6). This is obvious, because we have used the same argument here as in the discussion leading to Eq. (5.6). We can rewrite Eq. (5.11) as

$$T_A = \frac{1}{\Omega_A} \int_{4\pi} T_b\, (\theta, \phi)\; g(\theta, \phi)\; d\Omega. \tag{5.14}$$

Thus the measured antenna temperature is the integral over the brightness temperature distribution over the celestial sphere weighted by the antenna pattern and normalised by the total antenna beam pattern solid angle.

We now consider sources of a finite angular extent with solid angle Ω_s, a frequency dependent flux density S_ν, expressed in jansky (1 Jy = 10^{-26} Wm^{-2} Hz^{-1}) and a brightness temperature T_b (in K) at wavelength λ. The flux density is defined by the following expression

$$S_\nu = \frac{2k}{\lambda^2} \int_{\Omega_s} T_b\, d\Omega. \tag{5.15}$$

Scanning the antenna beam across a source with a brightness distribution $T_b(x', y')$, the antenna temperature $T_A(x, y)$ is given by the *antenna convolution integral*

$$T_A(x, y) = \frac{1}{\Omega_A} \int\int g(x - x', y - y')\, T_b(x', y')\, dx'\, dy' . \qquad (5.16)$$

The measured distribution T_A is the true distribution T_b"smeared" by the finite width of the beam. For simple functional forms of T_b, for instance a gaussian or disc distribution, the antenna pattern $g(x, y)$ may be found in closed form from Eq. (5.16). In the extreme case that the source size is much smaller than the beam width (we call this a "*point source*", a delta function), the convolution reduces to the integral over the antenna pattern, which is reproduced in the shape of the measured function T_A. From Eqs. (5.6, 5.14 and 5.15) we can easily derive the following expression for a point source

$$S_\nu A = 2 k T_A . \qquad (5.17)$$

Thus the observation of a **point source** with known flux density delivers immediately the absorption area of the antenna. Alternatively, the observation of a point source with an antenna of known absorption area yields the flux density of the source. For sources of finite angular extent the situation is less simple, because the source "fills" a part of the antenna beam and the recovery of the true brightness distribution of the source from Eq. (5.16) is more complicated. We discuss this in the next section.

■ 5.3. Efficiencies and Corrections for finite source size

5.3.1. Aperture and Beam Efficiency

We now define two quantities which describe the efficiency with which the antenna collects the received radiation. First, a basic quantity, already introduced in Ch. 4.1, is the *aperture efficiency*, defined as

$$\eta_A = A \, / \, A_g , \qquad (5.18)$$

where A is the maximum absorption area and A_g the geometrical area of the antenna aperture. It indicates the efficiency with which the radiation from a *point source* is collected. It is directly related to the gain G, as introduced in Ch. 5.1, which is often the primary antenna parameter used by communication engineers. The aperture efficiency can be separated into a number of different components, which we have qualitatively described in Ch. 4.1. A knowledge of the aperture efficiency is required at all wavelengths used to obtain quantitative data on the intensity of the source received. Knowing η_A over a wide frequency range also enables us to draw interesting conclusions about the antenna performance. We shall give examples later. Often, the source of radiation cannot be considered point-like and a different approach to determining the antenna efficiency must be taken.

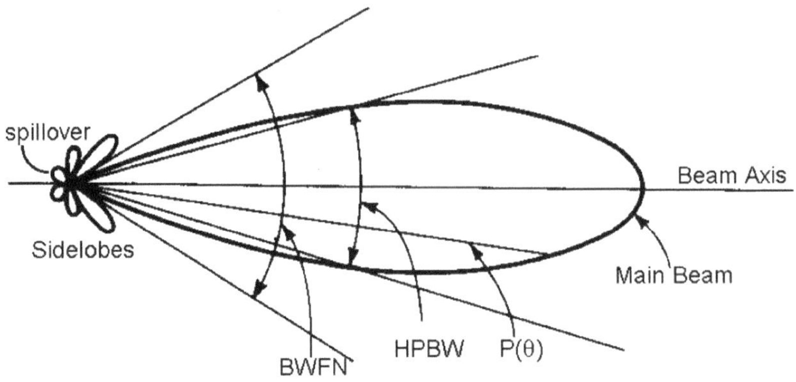

Fig. 5.2. Sketch in polar coordinates of the antenna beam with main beam and sidelobes. The half-power beam width (HPBW) and beam width to first null (BWFN) are indicated as well as the general power level P(θ).

We can separate the antenna beam solid angle into two components (Fig. 5.2), one containing the complete main beam Ω_m and the other Ω_l representing all side- and back-lobes:

$$\Omega_A = \Omega_m + \Omega_l = \iint_{\text{main beam}} g(\theta, \phi)\, d\Omega + \iint_{\text{sidelobes}} g(\theta, \phi)\, d\Omega. \quad (5.19)$$

A useful quantity for the measurement of extended sources is the "*main beam efficiency*" (often loosely called "beam efficiency"), defined as

$$\eta_{\text{mB}} = \Omega_m / \Omega_A, \quad (5.20)$$

where Ω_m is the *main beam solid angle*. Thus the beam efficiency is the fraction of all power received which enters the main beam (assuming that the antenna is surrounded by a source of uniform temperature). Normally the main beam is defined to extend to the first null in the radiation pattern. Sometimes it is convenient to include also sidelobes covering the source under study into the "extended main beam efficiency", which we shall denote η_B. We discuss examples later. Using the relations of Eq. (5.7) and the above definitions of the aperture and beam efficiency we can derive the following useful expression

$$\eta_{\text{mB}} = \frac{\Omega_m A_g}{\lambda^2} \eta_A . \quad (5.21)$$

Thus, if we determine η_A from a measurement of a point source with known flux density using Eq. (5.17) and we determine Ω_m from the measured beam scan over that source, we find the main beam efficiency η_{mB}. We shall see below that it is also possible to measure the beam efficiency directly using a source of known, finite angular extent.

5.3.2. Convolution of the beam with a source of finite size

We continue with the description of the interaction of the antenna beam with a radiating source of finite angular extent. It will be clear that radio sources with an angular size which is not negligible with respect to the antenna HPBW, cause some complication in the analysis. On the other hand, measurements on sources with different angular size can be helpful for the derivation of certain antenna parameters. This is caused by the fact that the sources subtend different portions of the antenna pattern.

Assume that we observe a source with the usual "on-off" technique, in which we place the beam on the center of the source and measure its intensity with respect to a measurement on a neighbouring spot of the sky, believed to be free of sources. We thus measure the product (rather than the convolution mentioned in Eq. (5.16)) of the antenna pattern and the source brightness distribution within the solid angle of the source Ω_s. For the moment we assume that $\Omega_s < \Omega_m$, the main beam solid angle. In analogy to Eq. (5.14) and using Eq. (5.7) we find the measured antenna temperature

$$T_A = \frac{A}{\lambda^2} T_b \int_{\mathrm{source}} \psi(\theta, \phi)\, g(\theta, \phi)\, d\Omega, \tag{5.22}$$

where we have introduced the normalised source brightness distribution $\psi(\theta, \phi)$. The source solid angle is the integral over the normalised source brightness distribution:

$$\Omega_S = \int_{\mathrm{source}} \psi(\theta, \phi)\, d\Omega. \tag{5.23}$$

We now substitute Eq. (5.15) into Eq. (5.22) and obtain

$$T_A = \frac{SA}{2k} \frac{1}{\Omega_S} \int_{\mathrm{source}} \psi(\theta, \phi)\, g(\theta, \phi)\, d\Omega = \frac{SA}{2k} \frac{\Omega_\Sigma}{\Omega_S}\ . \tag{5.24}$$

where we have introduced the *beam-weighted source solid angle*

$$\Omega_\Sigma = \int_{\mathrm{source}} g(\theta, \phi) \,.\, \psi(\theta, \phi)\, d\Omega\, . \tag{5.25}$$

The factor $K \equiv \frac{\Omega_S}{\Omega_\Sigma}$ corrects the measured antenna temperature for the weighting of the source brightness distribution by the antenna beam. Thus, the flux density of a source of finite extent and smaller than the antenna beam can now be written as

$$S = \frac{2k}{A} K\, T_A. \tag{5.26}$$

Note that this expression implicitly contains the most basic antenna characteristic, the aperture efficiency $\eta_A = A \,/\, A_g$ (Eq. (5.18). It can normally be accurately determined from the observation of a point source with known flux density or a small source, like a planet, where we can calculate K reliably. Thus, as long as the source is smaller than the beam, there is no need to invoke the beam efficiency, which is more difficult to establish, because the entire main beam must be measured. With contemporary, large millimeter wavelength telescopes with beamwidths of the order of 10 arcseconds, this is often not the case and the aperture efficiency must be determined via the measurement of the beam efficiency.

The brightness distribution of many radio sources can be represented by either a gaussian distribution with source half-power width θ_s or a disc distribution with angular diameter θ_d. Thus it is convenient to describe the main beam also by a gaussian function. Despite the fact that the theoretical antenna pattern of a tapered circular aperture distribution has the shape of Lambda-functions (see Ch. 3), often the main beam, down to a measurement level of about -20 dB, can be described quite accurately by a gaussian function (see the box in Sec.5.7). Assuming for simplicity that the beam is circularly symmetric, the normalised beam pattern can then be written in the usual form as

$$g(\theta,\ \phi) = \exp\left[-\frac{\theta^2 + \phi^2}{2\sigma^2}\right],$$

where θ and ϕ indicate the angles in the two principal (electric and magnetic) planes of the beam. For convenience we want to express the standard deviation σ in terms of the full half-power beamwidth θ_A. Thus we have $\exp\left[-\frac{(0.5\,\theta_A)^2}{2\sigma^2}\right] = 0.5$ from which we readily find $\sqrt{2\sigma^2} = \theta_A \,/\, 2\sqrt{\ln 2} = 0.6006\,\theta_A$. The Gaussian expression for the beam is now

$$g(\theta,\ \phi) = \exp\left[-\frac{4\ln 2\,(\theta^2 + \phi^2)}{\theta_A{}^2}\right] = \exp\left[-\frac{\theta^2 + \phi^2}{(0.6006\,\theta_A)^2}\right], \tag{5.27}$$

where θ_A is the full half-power beamwidth (HPBW). Integrating this function over the solid angle of the beam (formally extending the integration to infinity) delivers the main beam solid angle for the gaussian approximation

$$\Omega_m = \int_{-\infty}^{\infty}\int_{-\infty}^{\infty} g(\theta,\phi)\,d\theta\,d\phi = \left(\tfrac{1}{2}\sqrt{\tfrac{\pi}{\ln 2}}\;\theta_A\right)^2 = 1.133\;\theta_A{}^2. \qquad (5.28)$$

A similar formula is valid for the symmetrical gaussian source distribution,

$$\Omega_s = 1.133\,\theta_s{}^2,$$

while the circularly symmetric disc distribution is represented by the trivial function

$$\psi(\theta) = 1 \;\text{ for }\; \theta < \theta_d \quad\text{and}\quad \psi(\theta) = 0 \;\text{ for }\; \theta > \theta_d$$

with

$$\Omega_d = \pi\,(\theta_d\,/\,2)^2 = 0.7854\,\theta_d{}^2.$$

We can now derive the factor $K \equiv \frac{\Omega_S}{\Omega_\Sigma}$ of Eq. (5.26) for the gaussian and disc source distributions, assuming the antenna beam to have a gaussian shape as well. For a symmetrical gaussian source Eq. (5.25) becomes

$$\Omega_\Sigma =$$

$$\int_{-\infty}^{\infty}\int_{-\infty}^{\infty} \exp\!\left[-\,4\ln 2\,\frac{\theta^2+\phi^2}{\theta_A{}^2}\right]\cdot\exp\!\left[-4\ln 2\,\frac{\theta^2+\phi^2}{\theta_s{}^2}\right] d\theta\,d\phi = \frac{\pi}{4\ln 2}\;\frac{\theta_s{}^2}{1+(\theta_s/\theta_A)^2}\,.$$

The equivalent relation for the circularly symmetric disc source is

$$\Omega_\Sigma = 2\pi\int_0^{\theta_d/2} \exp\!\left[-\,4\ln 2\,\frac{\theta^2}{\theta_A{}^2}\right]\theta\,d\theta = \frac{\pi}{4\ln 2}\;\theta_A{}^2\left\{1-\exp\!\left[-\ln 2\left(\frac{\theta_d}{\theta_A}\right)^2\right]\right\}.$$

With the relations for Ω_s and Ω_d, given above, we obtain for the factor K:

$$K = 1 + x^2, \qquad \text{with } x = \theta_s\,/\,\theta_A,\ \text{gaussian source} \qquad (5.29a)$$

$$K = \frac{(x/1.2)^2}{1-\exp\{-(x/1.2)^2\}}, \text{ with } x = \theta_d\,/\,\theta_A < 1,\ \text{disc source.} \qquad (5.29b)$$

The observed scan profile of a gaussian beam scanning over a gaussian source can be computed from the convolution integral, Eq. (5.16). We recall that the convolution of two functions is equal to the product of their Fourier transforms (e.g. Bracewell, 1965). Now, the Fourier transform of a gaussian is a gaussian and the product of two gaussians is again a gaussian. Thus we find that the measured half-power width θ_0 is given by the expression

$$\theta_0 = \sqrt{\theta_A{}^2 + \theta_s{}^2} \, . \tag{5.30}$$

For the disc distribution with $\theta_d < \theta_A$ (or x < 1 in Eq. (5.29b)) the result of the convolution is nearly gaussian and the measured half-width of the scan is found to be

$$\theta_0 = \sqrt{\theta_A{}^2 + \frac{\ln 2}{2}\, \theta_d{}^2} \, . \tag{5.31}$$

Table 5.1. Correction factors for measurements with extended sources

– x	K gaussian	K disc	θ_0 / θ_A gaussian	θ_0 / θ_A disc
0.0	1.000	1.000	1.000	1.000
0.05	1.0025	1.0009	1.0025	1.0013
0.1	1.010	1.0035	1.005	1.0017
0.2	1.040	1.0140	1.020	1.0069
0.3	1.090	1.0316	1.044	1.0155
0.5	1.250	1.0893	1.118	1.0424
0.7	1.490	1.1798	1.221	1.0816
1.0	2.0	1.3871	1.414	1.1604

Table 5.1 contains the numerical values for the correction factor K and the linear beam broadening for gaussian and disc distributions. The *Mathematica* routine [Mat.5.1] produces Fig. 5.3.

For larger disc sources (x > 1), the convolution (Eq. 5.16) does not produce a simple formula for the measured half-power width. P. Stumpff (unpublished) has numerically calculated this case and his results are shown in Fig. 5.4 for values of 1 to 5 for the ratio of the source diameter to antenna beamwidth. Note that for some cases the measured scan width is slightly smaller than the true source width.

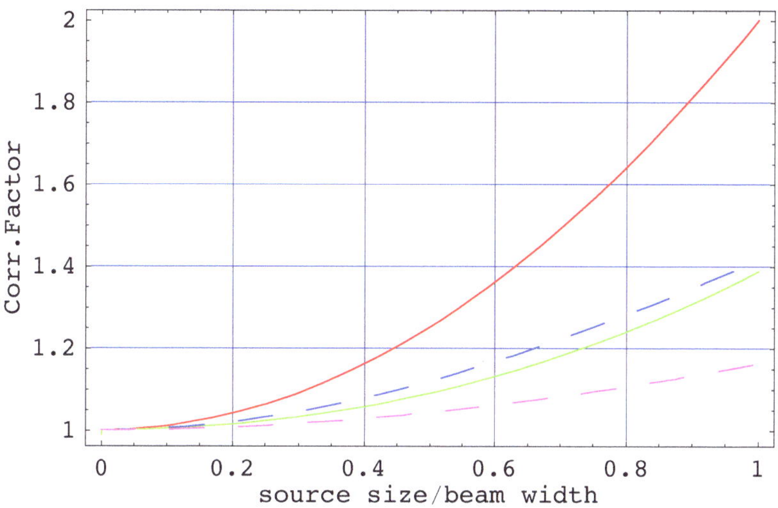

Fig. 5.3. Correction factors for measurements with extended sources. Red and green curves give the ratio of the true versus measured antenna temperature for gaussian and disc source distribution. Blue and magenta dashed curves show ratio of measured to true antenna HPBW for gaussian and disc, respectively.

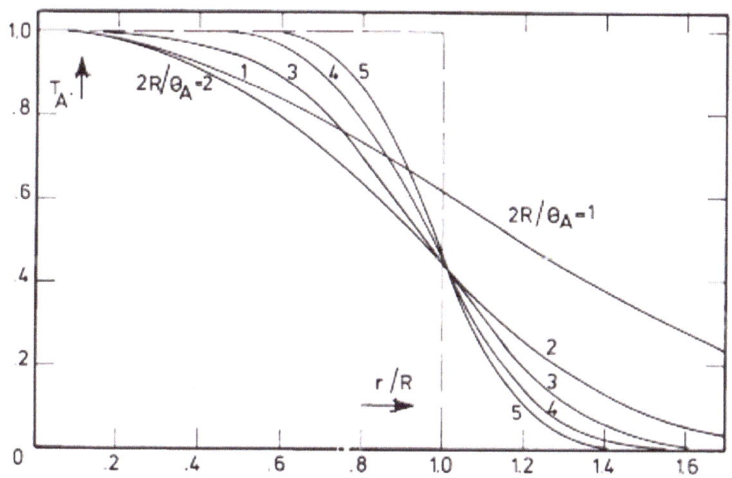

Fig. 5.4. Normalised antenna temperature versus normalised scan angle of an antenna beam with width θ_A over a disc source of radius R with ratio 2 R/θ_A as parameter. (P. Stumpff, unpublished)

In the case of a very large disc where $\theta_d / \theta_A >> 1$ the problem can be reduced to the one dimensional convolution of a gaussian with an essentially straight edge. The gaussian beam can be separated into the product of two one-dimensional gaussians, while the brightness of the source is approximated by $\psi(x, y) = H(x)$, this being the Heaviside step function in x at the edge of the source. The convolution integral (Eq. (5.16)) is simplified to

$$T_A(x) \propto T_b \int_{-\infty}^{\infty} g(y-y')\,dy' \int_{-\infty}^{\infty} g(x-x')\,H(x')\,dx'. \tag{5.32a}$$

The first integral reduces to a constant. Differentiating the equation leads to

$$\frac{dT_A}{dx} \propto T_b\,C \int_{-\infty}^{\infty} g(x-x')\,\frac{d}{dx}\,H(x')\,dx' = C\,T_b\,g(x), \tag{5.32b}$$

when we consider that the derivative of the Heaviside function is the Dirac delta function $\delta(x-x')$ and the convolution $g(x') * \delta(x-x') = g(x)$. Thus the antenna beam shape can be recovered quite reliably (depending on the signal to noise ratio) by numerical differentiation of the observed scan across the edge of the large source.

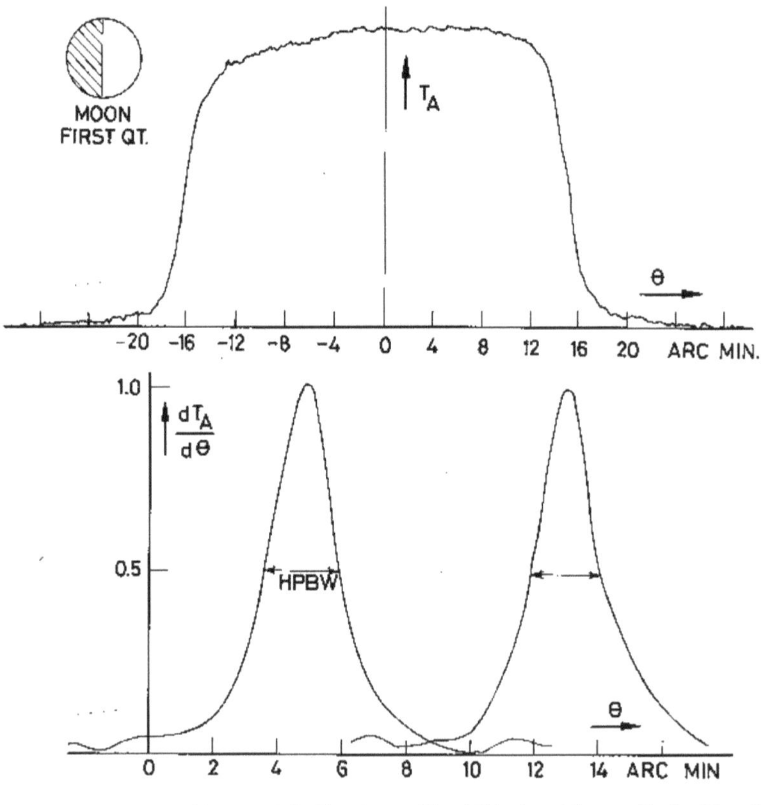

Fig. 5.5. Scan of the moon (diameter 30') with a beamwidth of 2' (top) and the graphically differentiated steep edges of the scan (bottom). These give a reasonable indication of the beam with a derived beamwidth of 2.1'. Some sidelobe or error pattern features are visible in the derived beam shape. Note the slight asymmetry in the Moon scan due to a temperature gradient across the surface (see Moon phase as indicated) at the wavelength of 2 cm.

This has been done successfully with beam widths of several arcminutes and the Moon or Sun as source (angular diameter 30 arcminutes). An example is shown in Fig. 5.5.

To illustrate this situation further, we write the integral of Eq. (5.32a) in *Mathematica* [Mat.5.2] where we first assume a beam of gaussian form with HPBW θ. The convolution with the straight edge (step function) delivers an Erf-function (see e.g. Abramowitz and Stegun, 1964, p. 295). Differentiating this result recovers the input gaussian (Fig. 5.6, red lines). Next the convolution with the beam in the form of the Lambda function is computed (Fig. 5.6, blue lines). Here we see the influence of the sidelobes, which are absent in the gaussian representation. Again the input Lambda function is recovered upon differentiating the convolution result. Note that to improve the visibility we plot here the convolution with the field pattern (not the usual power pattern), which shows negative sidelobes.

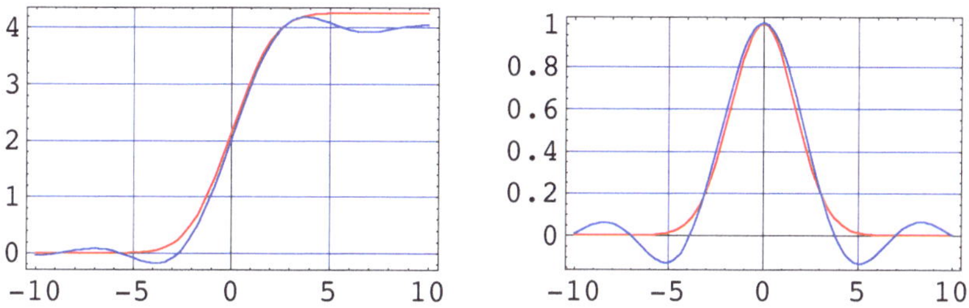

Fig. 5.6. The convolution of a gaussian (red) and a Lambda-function (blue) with a straight edge (left) and the original beams, obtained from differentiating the convolution results (right).

In the section with Mathematica routines at the end of this Chapter, the reader will find a quantitative comparison of the gaussian and lambda function representation of the antenna beam [Mat.5.3].

5.3.3. Beam efficiency and intensity calibration

We now return to the discussion of the beam efficiency. Recall the expression for the main beam efficiency (Eq. (5.21)):

$$\eta_{\mathrm{mB}} = \frac{\Omega_m A_g}{\lambda^2} \eta_A .$$

If the beam shape is known, the beam solid angle Ω_m can be computed and we obtain a value for the beam efficiency η_{mB}. Using the gaussian approximation of Eq. (5.28) for the beam solid angle and remembering that $\theta_A = b\,(\lambda/d)$ (see Ch. 4.2.2, Eq(4.13)), we obtain

$$\eta_{\mathrm{mB}} = \frac{\pi d^2}{4} \frac{1.133}{\lambda^2} \left(\frac{b\lambda}{d}\right)^2 \eta_A = 0.89\, b^2\, \eta_A . \qquad (5.33)$$

For a typical aperture illumination with -12 dB edge taper, we have b=1.16 and hence $\eta_{\mathrm{mB}} = 1.2\,\eta_A$.

The measured antenna temperature as function of the source solid angle for a certain antenna beamwidth is qualitatively illustrated in Fig. 5.7. As long as $\Omega_s << \Omega_m$ the beam pattern is essentially unity over the source area (assumed uniformly bright) and the antenna temperature increases linearly with increasing source solid angle. At some point the weighting by the beam becomes significant and the antenna temperature increases proportional to the weighted source solid angle until it reaches $T_A = \eta_{\mathrm{mB}}\, T_b$ when $\Omega_\Sigma = \Omega_m$. For, if we combine Eq. (5.26) and Eq. (5.21) with $\Omega_\Sigma = \Omega_m$, we obtain

$$S = \frac{2k}{\lambda^2}\,\frac{1}{\eta_{\mathrm{mB}}}\,T_A\,K\,\Omega_m = \frac{2k}{\lambda^2}\,\frac{1}{\eta_{\mathrm{mB}}}\,T_A\,\Omega_s = \frac{2k}{\lambda^2}\,T_b\,\Omega_s ==> T_A = \eta_{\mathrm{mB}}\,T_B.$$

If the antenna were free of sidelobes, a further increase in source size would not lead to an increase in T_A. In reality the sidelobes will add signal until T_A reaches T_b when the source completely surrounds the antenna.

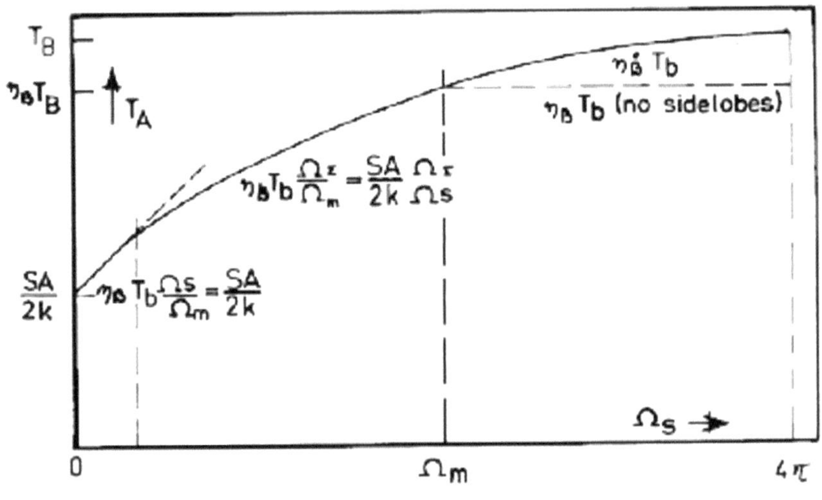

Fig. 5.7. Illustrating the run of measured antenna temperature T_A as function of the source solid angle Ω_s. Note that η_B in this figure is denoted η_{mB} in the text.

This suggest a way to treat the case where the source is significantly larger than the mainbeam. In this case a number of sidelobes and the error pattern (see Ch. 4.6) will receive power from the source and contribute to the measured antenna temperature. Although the level of these beam features is low, the angular size is appreciable and in case of observations of giant molecular clouds in our Galaxy with the narrow beam of a large millimeter telescope the power received in the sidelobes can be considerable. For an accurate measurement, both the detailed form of the beam and sidelobes and the brightness distribution of the source would need to be known. Normally neither are. As a practical approach we define an *effective beam efficiency* η_B (thus not restricted to the mainbeam) as

$$\eta_B(\Theta) = \frac{1}{\Omega_A} \int_{\Theta_S} g(\theta, \phi)\, d\Omega, \tag{5.34}$$

in which the integration is extended over a solid angle Θ_S, equal to that of the source. For a known source size, Eq. (5.34) would be the best representation of the coupling of the extended antenna pattern to the source. For a uniformly bright source of size Θ_S, the measured antenna temperature would now be $T_A = \eta_B T_B$ (Gordon et al, 1992).

We can obtain an impression of the effective beam efficiency as function of the angle θ by measuring sources of different angular size, as the planets (a few arcseconds to one arcminute) and the Moon (30 arcmin) and interpolate in between. We shall illustrate below how this can be used to obtain an indication of the sidelobe level of the antenna. The interpolation from 1 to 30 arcminutes is of course prone to error, but it might still be better than trying to make a full theoretical calculation of the effective beam efficiency. For very large objects, beyond 30' size, we might include the "forward beam efficiency" (over 2π steradian) in the interpolation. This is sometimes determined from the measurement of the atmospheric emission over the full 90° elevation range - the so-called "sky-dip" (Kutner and Ulich, 1981). Its value will be very close to one.

At millimeter wavelengths, especially in the observation of large molecular clouds, procedures have been proposed by Kutner and Ulich (1981), which describe observations in terms of the parameter T_R^*, the observed antenna temperature corrected for all telescope-dependent parameters *except* the "coupling" of the antenna to the source brightness distribution. Under the assumption of a uniformly bright source, we find that the effective beam efficiency, defined above is related to these "coupling" (η_c) and "extended source" (η_s) efficiencies of Kutner and Ulich by the relation $\eta_B = \eta_c\, \eta_s$. Unfortunately, η_c generally cannot be measured and can be calculated only under simplifying assumptions. It appears preferable to use the η_B, introduced above in these cases. If a reasonable estimate of $\eta_B(\Theta)$ for a source of solid angle Θ is available, one could correct the measured antenna temperature at each point of the map into a "main beam" value by multiplying by η_{mB} / η_B. The intensities of the resulting map would then appear as to have been observed with a "clean" beam of efficiency η_{mB}.

A useful summary of beam efficiency measurements, along with practical data on planetary brightness temperatures and applied to the Caltech Submillimeter Telescope, is presented by Mangum (1993).

■ 5.4. Sidelobe level and error pattern

5.4.1. Diffraction beam sidelobes

The availability of sources over a range of angular size, from real "point-like" to significantly larger than the antenna beamwidth, can be used to obtain an estimate of the sidelobe level near the main beam without the need to measure these individually, which would require a very high signal to noise ratio. By the same token, it is possible to estimate the level of the "error pattern" in the case of significant random errors in the reflector surface profile.

The radiation pattern of a perfect reflector consists of a central lobe surrounded by ring shaped sidelobes of decreasing amplitude (the Airy pattern, see Ch. 3). Assuming circular symmetry, we represent the pattern, as function of the radial angular coordinate θ, by

$$g(\theta) = g_m(\theta) + \textstyle\sum_i g_{\mathrm{li}}(\theta) .$$

From the representation of the beam in terms of the Lambda function (see Ch. 3), we can derive the following reasonably good approximation for the half-power width θ_{li} and the radius of maximum intensity r_i of the ith sidelobe as $\theta_{\mathrm{li}} = \theta_A / 2$ and $r_i = \theta_A\,(\frac{1}{2} + i)$. Using again the gaussian approximation for the main beam, we can now write for the pattern solid angle up to and including i sidelobes

$$\Omega = \Omega_m + \textstyle\sum_i \Omega_{\mathrm{li}} =$$
$$1.133\,\theta_A{}^2 + \textstyle\sum_i \pi\,\theta_A{}^2 (\frac{1}{2} + i)\; g_{\mathrm{li}}(\max) = \Omega_m \{1 + \frac{\pi}{1.133}\,(\frac{3}{2}\,g_1 + \frac{5}{2}\,g_2 + ..)\}$$

We now apply the concept of *effective beam efficiency* η_B, as introduced in Eq. (5.34). Thus by measuring η_B on an extended source and η_{mB} on a point source, we obtain an estimate of the average level of those sidelobes covering the source from the relation

$$\frac{\eta_B}{\eta_{\mathrm{mB}}} = 1 + 2.77\,(\tfrac{3}{2}\,g_1 + \tfrac{5}{2}\,g_2 + ..). \tag{5.35}$$

We can also use the curve of Fig. 3.5, based on the Bessel function representation of the radiation pattern. From that curve and the numbers in Table 3.1 we derive values for the ratio $\eta_B / \eta_{\mathrm{mB}}$ which are very close to those found above. Remember that we assume a uniform brightness distribution over the source of a known size. If we take as an example a 30 m diameter telescope operating at 1.2 mm wavelength, we have a HPBW = 10 arcseconds. With planets in size between a few arcseconds (Uranus, Neptune) and 1 arcminute (Jupiter, Venus) this method can be used to estimate the average near-in sidelobe level.

5.4.2. Error pattern due to random surface errors

In the case, where the sidelobe level is dominated by the scattering from random errors in the reflector profile (the so-called "Ruze-error"), we can use the concept of "effective" beam efficiency to estimate the size and level of the error pattern. We have already treated the theory of these errors and the resulting radiation pattern characteristics in Chapter 4.6. There we described the beam pattern as the sum of the *diffraction pattern* g_D of the perfect reflector and the *error pattern* g_E, caused by the scattered radiation from the random errors. The errors cause an rms phase fluctuation over the aperture of

$$\sigma = 4\,\pi\,\epsilon\,/\,\lambda,$$

where the surface error ϵ can be weighted by the aperture illumination function, if desired. Also, we repeat that ϵ is the deviation parallel to the reflector axis, i.e. half the total pathlength error. It is related by Eq. (4.43) to the error normal to the reflector surface, which is the quantity usually measured or calculated from structural analysis.

We repeat Eq. (4.42) for the relative change in aperture efficiency caused by the random errors,

$$\frac{\eta_A}{\eta_{A0}} = \exp(-\sigma^2) + \frac{1}{\eta_{A0}}\left(\frac{c}{d}\right)^2 \{1 - \exp(-\sigma^2)\}. \tag{5.36}$$

Normally the second term has a negligible contribution to the aperture efficiency and can be dropped for this purpose, leading to the widely known "Ruze" formula for the loss of gain due to random surface errors. If we measure η_A at a number of sufficiently separated wavelengths, we can make a good estimate of both η_{A0} and ϵ. This is illustrated in Fig. 5.8, which shows the logarithm of the measured aperture efficiency as function of the reciprocal square of the wavelength [Mat.5.4]. The slope of the fitting line gives the rms surface error from $<\text{slope}> = (4\,\pi\,\epsilon)^2$ and the intercept

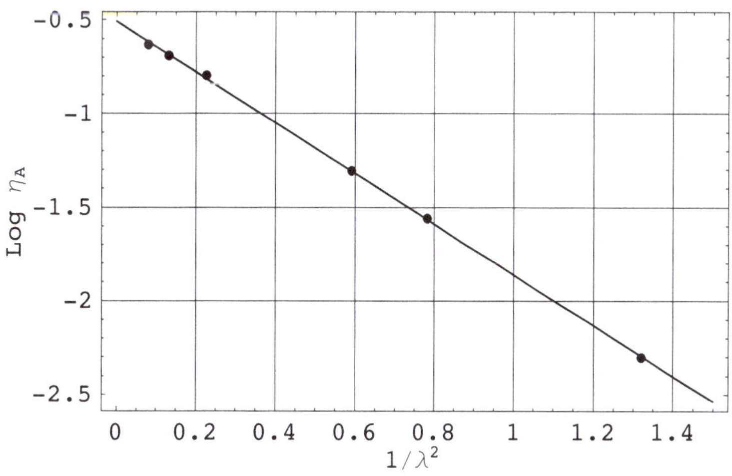

Fig. 5.8. Measured aperture efficiency at 6 wavelengths. Plotted is the logarithm of the efficiency against the reciprocal square of the wavelength. The slope gives the rms error as 92 μm.

for λ->∞ delivers η_{A0}. These are measurements made early in the operation of the IRAM 30-m millimeter telescope. The derived surface error is 92 μm, which is in reasonable agreement with a holographic surface measurement of 85 μm. Since then the surface has been significantly improved. The value $\eta_{A0} = 0.6$ agrees with the calculated value from the illumination function.

Also, a change in measured η_A as function of elevation angle can be used to compute the relative change in ϵ and give an impression of the gravitational deformations of the reflector surface, which can be compared with structural finite element analysis. An example is shown in Fig. 5.9, based on efficiency measurements with the IRAM 30-m mm-telescope at 1.2 mm wavelength, shown as the solid line. The dashed line is the predicted change in efficiency from the finite element analysis. On the right hand axis is a scale with the inferred increase in the reflector rms surface deviation. At 80° elevation angle the predicted deformation is 40±10 μm, while the measurement indicates 52±5 μm. It should be noted that it is essential to accurately correct for the considerable atmospheric attenuation at these sort wavelengths as function of elevation, not a simple task because of the variable nature of the troposphere.

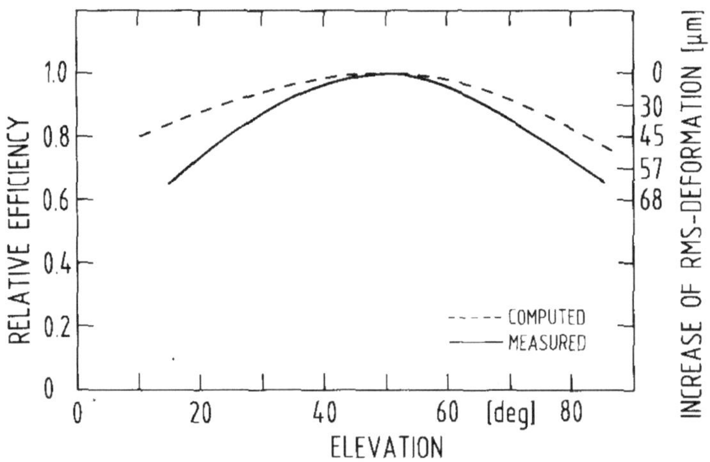

Fig. 5.9. Calculated (dashed) and measured (full) change in aperture efficiency, and hence increase in reflector rms surface deviation (right hand scale) as function of elevation angle. In view of the uncertainties, the results are consistent.

If a measurement has sufficient signal to noise ratio to determine the error pattern reliably, we can derive the correlation length **c** from the ratio of the peaks of the error pattern and the diffraction pattern (see Eq. (4.44)

$$\frac{g_E(0)}{g_D(0)} = \frac{1}{\eta_{A0}} \left(\frac{c}{d}\right)^2 \{\exp(\sigma^2) - 1\}, \qquad (5.37)$$

while the gaussian shaped error pattern has a half-power width

$$\theta_E = 2\sqrt{\ln 2}\,\left(\frac{2\lambda}{\pi c}\right) = 1.06\,\frac{\lambda}{c}\,. \tag{5.38}$$

The HPBW of a tapered aperture distribution with about -12 dB edge illumination is $\theta_A \simeq 1.12\,\lambda/d$ (see Ch. 4.2.2) and hence

$$\theta_E\,/\,\theta_A \simeq 0.94\,d\,/\,c\,. \tag{5.39}$$

Although the error pattern is weak with respect to the main beam in most cases, its large width causes it to contain a significant part of the received power. In particular in observing extended objects, this can lead to erroneous conclusions as to the brightness temperature of the source.

We have already illustrated above the importance of a knowledge of the antenna sidelobe structure for the correct interpretation of certain astronomical observations. On the other hand, often new observational results may be expected by using the telescope at the limits of its capability. An experimental determination of the error pattern will help in deciding the usefulness of the antenna for the observational goal. We illustrate this with an example of a measurement of the beam pattern, including

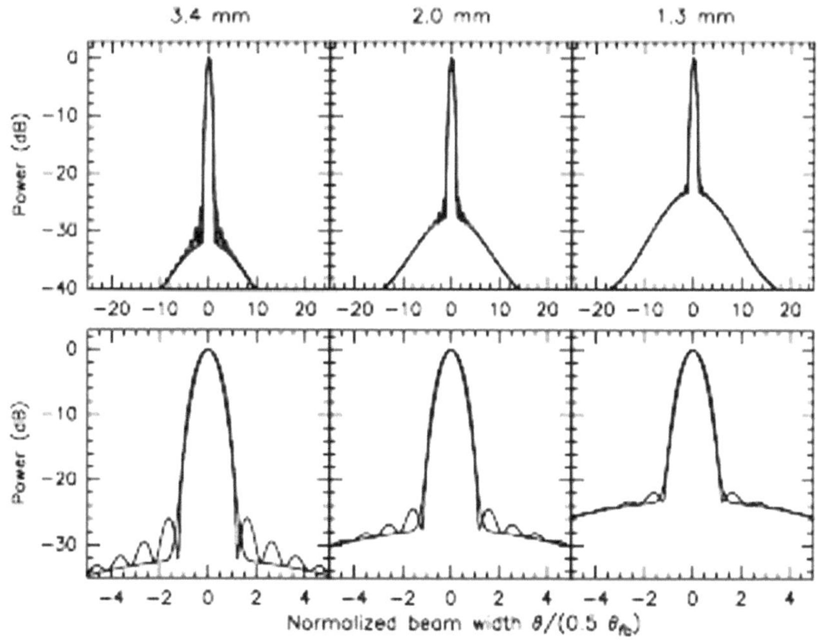

Fig. 5.10. Beam patterns of the IRAM 30-m telescope derived from differentiated Moon scans at 3.4, 2.0 and 1.3 mm wavelength. The lower section shows the central part at a larger angular scale. Next to the diffraction (main) beam two errors patterns can be seen, which have a width of 17 and 90 times the diffraction beamwidth, respectively.

the error beam, at several wavelengths with the IRAM 30-m diameter millimeter telescope (Greve et al., 1998). The results are shown in Fig. 5.10. To obtain sufficient signal to noise ratio these data were obtained by differentiation of scans across the edge of the Moon, as discussed above. One can discern a main beam (the diffraction beam) and two error beams, which increase in central intensity with decreasing wavelength. In the upper part of the figure the very wide error beam is obvious. In the lower section of the figure only the central part of the composite beam is shown and a second, narrower error beam can be seen, upon which the diffraction beam with its sidelobes is superposed. From the analysis of these measurements, using the formulae presented above, one can deduce a first, narrower error beam with a width of 17 times the diffraction beam width, leading to a correlation length of 1.5 - 2.0 m. This can be identified with the surface panel frames with a typical length of 2 m and width of 1 - 1.5 m. These frames are supported on the backup structure on four points at their corners. The second, broader error beam has a width of about 90 times the diffraction beam, which gives a correlation length of 0.3 - 0.5 m. This is representative for small deformations within the panels, which are supported at about this interval on the panel frames. Such accurate measurements are not only useful for an understanding of the behaviour of the antenna; they are also essential in the correction of observations of sources of large angular extent.

Finally we illustrate how an observation of both a small and an extended source can be used to estimate the correlation length of the surface errors without making a full scan of the antenna pattern to very low power levels, as done in the example above. This assumes that the error pattern level is stronger than the intrinsic sidelobe level of the perfect antenna, i.e. at about -20 dB or stronger. Again we use the effective beam efficiency η_B, for a source width θ_s (with $\theta_A < \theta_s < \theta_E$) written as

$$\eta_B = \frac{1}{\Omega_0} \exp(-\sigma^2) \int_{\Omega_s} (g_D + g_E)\, d\Omega \,,$$

where Ω_0 is the antenna solid angle of the perfect antenna. The integral over the diffraction beam g_D delivers $\eta_{\rm mB}$ when measured on a small source, where the contribution of the error beam is negligible. Integrating the error beam g_E, as given in Eq. (4.40a), over the source solid angle, we derive

$$\eta_B - \eta_{\rm mB} = \{1 - \exp(-\sigma^2)\} \left\{1 - \left[\exp - \left(\frac{\pi c \theta_s}{4\lambda}\right)^2\right]\right\}.$$

Writing $\eta_{\rm B0}$ for the beam efficiency of the perfect reflector, which we find from $\eta_{\rm A0}$ with Eq. (5.21), we obtain $\eta_{\rm B0} - \eta_{\rm mB} = 1 - \exp(-\sigma^2)$ and combining this with the preceding expression finally yields

$$\frac{\eta_B - \eta_{\rm mB}}{\eta_{\rm B0} - \eta_{\rm mB}} = 1 - \exp\left\{-\left(\frac{\pi c \theta_s}{4\lambda}\right)^2\right\}, \qquad (5.40)$$

from which it is easy to solve for the correlation length c

$$c = \frac{4\lambda}{\pi\,\theta_s}\left\{-\ln\left(\frac{\eta_B - \eta_{\mathrm{mB}}}{\eta_{\mathrm{B0}} - \eta_{\mathrm{mB}}}\right)\right\}^{0.5}. \tag{5.41}$$

As we saw in the earlier example, the correlation length is often determined by the size of the individual panels, constituting the reflector. Thus values of d/c will typically lie between 6 and 20, depending on the number of panel rings. An example of the use of Eq. (5.41) is shown in Table 5.2.

Table 5.2. Measured parameter of a 25 m radio telescope

λ (cm)	η_A	η_{mB}	η_B
49.2	0.59	0.79	–
21.2	0.54	0.73	–
6.0	0.48	0.70	–
2.9	0.38	0.53	0.70

These are measurements of a 25 m Westerbork SRT antenna (Baars et al., 1973). Extrapolating the measured values to $\lambda \to \infty$ we find $\eta_{A0} = 0.60$ and $\eta_{B0} = 0.80$. With the measured effective beam efficiency on the Moon of 0.70 at 2.9 cm and its angular size $\theta_s = 30'$ we find from Eq. (5.41) that c = 4 m, which is in good agreement with the average size of the surface panels. Also from these data we derive with Eq. (5.36) (see also Fig. 5.8) the rms surface error of $\epsilon = 1.5$ mm, which compares well with the directly measured value of 1.4 mm.

■ 5.5. Pointing and focus corrections and optimisation

For an optimal performance of the reflector antenna, it is important that the system is properly focussed. For a prime focus reflector this means localizing the phase center of the feed in the focus of the primary paraboloid. In a Cassegrain system, the most important criterion is the coincidence of one of the foci of the hyperboloidal secondary reflector with the primary focus of the paraboloid. The location of the feed in the secondary focus is far less critical, as we have seen in Chapter 4. In this section we discuss the influence of a defocus, either axial or lateral, on the pointing characteristics. In addition we describe the method to establish the overall pointing model of the antenna.

5.5.1. Pointing aspects of defocus

The formulae for the change in pointing direction, as function of the lateral shift and rotation of the feed, main reflector and secondary, are presented in Table 5.3, Eq. (5.42). The geometry is illustrated in Fig. 5.11. Translation is reckoned positive in the positive y-direction and rotation is positive when right-handed with respect to the z-axis. Using the symbols K_p and K_s for the Beam Deviation Factor in the primary and secondary focus, respectively, we obtain the components of Table 5.3.

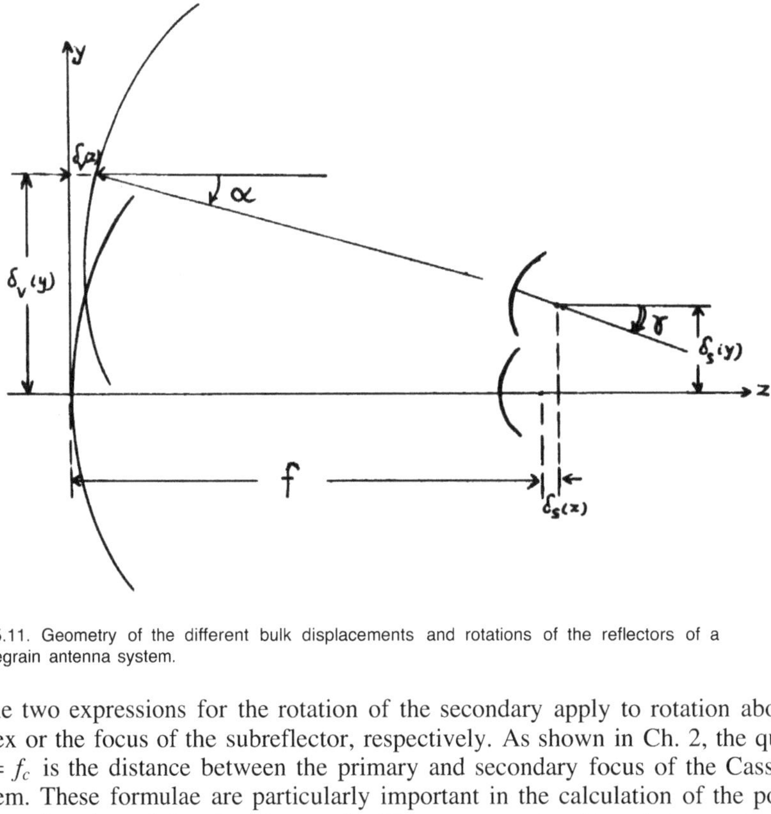

Fig. 5.11. Geometry of the different bulk displacements and rotations of the reflectors of a Cassegrain antenna system.

The two expressions for the rotation of the secondary apply to rotation about the vertex or the focus of the subreflector, respectively. As shown in Ch. 2, the quantity $2c = f_c$ is the distance between the primary and secondary focus of the Cassegrain system. These formulae are particularly important in the calculation of the pointing shift as function of the elevation angle. The components of translation and rotation are predicted from the Finite Element Analysis (FEA) of the structure. In an optimized design, the terms will cancel each other partially, leading to a relatively small pointing change with elevation.

5.5.2. General pointing model of the antenna

The basic method to point a telescope or antenna at a target of known position is to use the axis-encoder position read-outs as return signal to the servo control system until the differences between commanded and encoder position of both axes are zero.

Table 5.3. The components of pointing error due to deformation of the antenna

defocus component	symbol	pointing error	equation
translation primary	δ_p (y)	$-K_p \frac{\delta_p(y)}{f}$	(5.42 a)
rotation primary	α	$(1+K_p)\,\alpha$	(5.42 b)
translation secondary	$\delta_s(y)$	$\left(K_p - \frac{K_s}{M}\right) \frac{\delta_s(y)}{f}$	(5.42 c)
rotation secondary (vertex)	γ	$-\left(\frac{K_p + K_s}{M+1}\right) \frac{2c}{f}\,\gamma$	(5.42 d)
rotation secondary (focus)	γ	$-\frac{K_s}{M} \frac{2c}{f}\,\gamma$	(5.42 e)
translation feed in sec.focus	$\delta_f(y)$	$\frac{K_s}{M} \frac{\delta_f(y)}{f}$	(5.42 f)

In practice there will be a residual pointing error, i.e. a difference between the true viewing direction and that indicated by the encoders, caused by imperfections in the geometry and the finite structural stiffness of the antenna. These imperfections fall under the categories of misalignment of the axes, gravitational bending of the structure as function of elevation angle, errors in the zero point and linearity of the encoders and refraction by the earth's atmosphere. It will be necessary to apply corrections to the commanded position to ensure that the antenna beam is directed precisely in the desired direction. These corrections can be established by measuring the apparent position of a large number of sources with accurately known celestial position and distributed over the entire visible sky. From these observations we can determine the numerical values of the coefficients of the so-called *pointing model* of the antenna. The pointing model is created in the context of a pointing theory, where we seek to establish the analytical relationship between the true coordinates of the source, the target, and the read-out of the telescope encoders using a set of physically reasonable relations representing the known or expected geometrical and structural imperfections of the antenna system. Once we have set-up these relations, the coefficients of their terms can be found from the set of pointing observations by a least squares treatment. We call these parameters the *pointing constants*. They might vary over time due to aging effects in the structure, variable wind forces and diurnal or seasonal temperature variations. Thus for reliable antenna pointing it will be necessary to regularly check the constants for their best value.

In the following discussion we follow essentially the treatment by P. Stumpff (1972). We assume that the antenna has an altazimuth mounting, although this is not necessary for the theory to be discussed. It is most convenient to transform the celestial source coordinate right ascension (α) through the known local time t_l to hour angle (h = α - t_l) and thence hour angle and declination to apparent azimuth A and elevation (e) angle, whereby the local latitude (ϕ) must be known. Azimuth and elevation are given by the following equations

$$\cos e \cos A = -\sin\phi \cos\delta \cos h + \cos\phi \sin\delta$$
$$\cos e \sin A = -\cos\delta \sin h \qquad (5.43)$$
$$\sin e = \cos\phi \cos\delta \cos h + \sin\phi \sin\delta,$$

where azimuth A is reckoned from north through east (0° < A < 360°) and elevation e from horizon to zenith (0° < e < 90°). For a source with precisely known position and a telescope position of exactly known latitude and a perfect clock, making the encoder readings equal to the source (A, e)-values would put the antenna beam exactly on the source, provided the antenna coordinate system (A_i, e_i) is strictly orthogonal without rotation or offset errors with respect to the astronomical system (A, e). It is the deviation between the instrumental and celestial coordinate system which necessitates the establishment of the pointing model and application of pointing errors in actual practice. Although latitude and clock errors can be incorporated in the least squares solution of the pointing constants, we shall here assume the errors in these to be zero, similarly to the assumption of perfect source coordinates. Normally, these errors will indeed be negligible compared to the pointing errors to be established.

We are thus forced to apply pointing corrections ΔA and Δe to the source coordinates (A, e) to make the beam direction coincide with the source for the indicated encoder positions (A_i, e_i). The first order pointing theory to be described now establishes a functional relationship between the observed errors, the derived pointing constants and the basic coordinates. We write the corrections as "indicated minus commanded" values as follows:

$$\Delta A = A_i - A \quad \text{for azimuth, and}$$
$$\Delta e = e_i - e \quad \text{for elevation, respectively.} \qquad (5.44)$$

From the theory of the universal astronomical instrument (see e.g. K. Stumpff, 1955 or Smart, 1962) the following expressions, valid to first order, can be derived

$$\Delta A = A_0 + c_1 \sec e_i - c_2 \tan e_i - z_a \sin(A_i - A_a) \tan e$$
$$\Delta e = e_0 + b \cos e_i - r \cot e_i - z_a \cos(A_i - A_a), \qquad (5.45)$$

where the constants have the following meaning

A_0 and e_0 = zero point offset of azimuth and elevation encoder, respectively

b = the gravitational bending constant of the elevation section of the antenna

c_1 = collimation error of the beam (non-perpendicularity of elevation axis and beam)

c_2 = collimation error of mount (non-perpendicularity of azimuth and elevation axes)

z_a = zenith distance (= 90° - elevation) of azimuth axis (azimuth axis tilt)

A_a = azimuth of the azimuth axis

r = refraction constant.

Note that A_a and z_a can also be replaced by the angles between the azimuth axis and the north-south and east-west planes.

If we assume the errors to be small we may replace the "indicated" coordinates (A_i , e_i) in Eq. (5.45) by the "true" source coordinates (A, e). In order to point the antenna at the true position of the source, the indicated coordinates must be increased by the values of Eq. (5.45). We now introduce the following set of

"pointing constants" (5.46)

$P_1 = -c_1$	collimation of beam
$P_2 = -A_0$	azimuth encoder offset
$P_3 = c_2$	collimation of mount
$P_4 = -z_a \sin A_a$	azimuth axis offset(E-W)
$P_5 = z_a \cos A_a$	azimuth axis offset (N-S)
$P_6 = -e_0$	elevation encoder offset
$P_7 = -b$	gravitational bending
$P_8 = r$	refraction

With these we can write the pointing corrections as

$$\cos e\ \Delta A = P_1 + P_2 \cos e + P_3 \sin e + P_4 \sin e \cos A + P_5 \sin e \sin A$$

$$\Delta e = P_6 + P_7 \cos e - P_4 \sin A + P_5 \cos A + P_8 \cot e. \qquad (5.47)$$

Special studies have been made of the atmospheric refraction and as a result this term is well known (see Ch. 6.2.5). Therefore it normally is kept outside the pointing model calculation and applied separately. Thus there remain 7 parameters to be determined when an antenna needs to update its pointing model. As said before this is accomplished by measuring the differences ΔA and Δe for a large number of sources, distributed over the entire sky and solving by least squares methods the set of equations (5.47) for the seven constants. These equations with their proper values for the constants are then incorporated in the servo control software to apply the necessary correction to the demanded celestial position to force the beam to be pointed to that position.

The approach to establishing the pointing model, as illustrated above, is now an inherent part of every antenna and telescope control system. Rigorously speaking it is based on approximations which are valid for small errors and positions not too close to the zenith. For the most accurate work with current radio and optical telescopes, a refinement of the analysis might be required.

These are provided by the software package TPOINT, written by Patrick Wallace (1997) and marketed by Tpoint Software. This package is in use at several observatories, among them ALMA and the VLT of ESO. TPOINT is compatible with the somewhat differently structured pointing algorithms described in Wallace (2002). These are implemented as the proprietary software TCSpk, also available from Tpoint Software, which can be used to build antenna control systems in which the pointing analysis and control are closely linked. The ESA 35m deep space antenna in Western Australia (Fig. 1.4) uses both TPOINT and TCSpk, as do a number of optical/IR telescopes.

The notations used by Wallace are different from those used above. For ease of comparison we summarise in Table 5.4 the two correction systems. Note that we have listed in the table the ΔA correction terms, while in Eq. (5.47) $\Delta A \cdot \cos e$ is used. An example of the basic output of TPOINT is presented in Fig. 5.12. The distribution of the pointing sources over the sky is indicated in the lower right. The final scatter of the measurements around the nominal direction of the viewing angle is shown in the lower left. The upper two rows of scatter are the deviations of several parameters with respect to azimuth and elevation directions, while the distribution of deviations is shown in the middle lower panel.

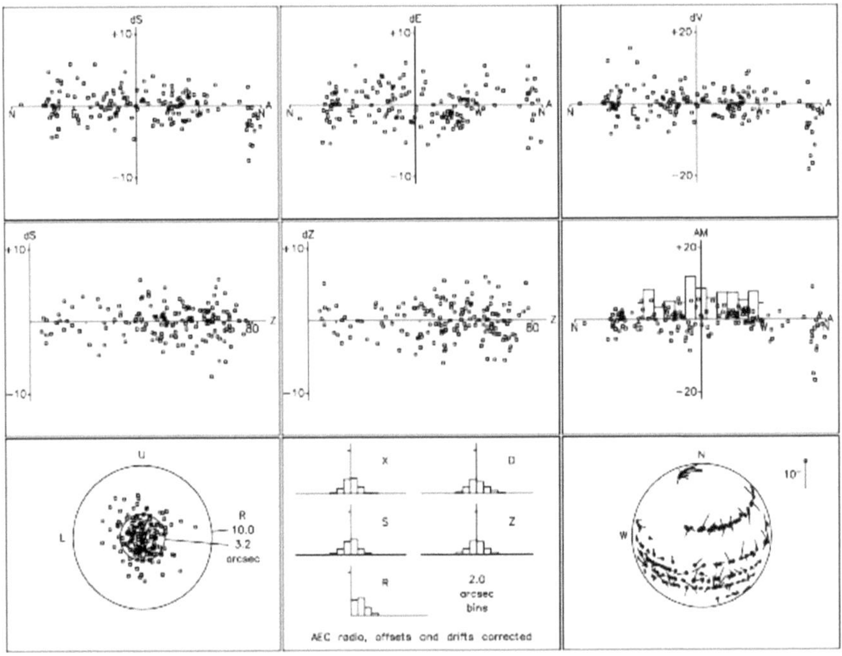

Fig. 5.12. Example of a standard output of the TPOINT program to evaluate the pointing parameters of an antenna; see text above figure. (after Wallace, 1997)

While the 7-parameter pointing model will account for the majority of the corrections, it is possible that the residuals after the least-squares fitting procedure exhibit characteristics which point to further systematic errors in the telescope system. Examples of this are harmonic terms caused by encoder and bearing run-out, azimuth bearing level variations due to azimuthal changes in the pedestal support stiffness, etc. These can be fitted to the appropriate functions and additional pointing model constants can be obtained and henceforth incorporated in the pointing system.

Table 5.4. Basic pointing model terms in the notation of Stumpff and Wallace

St	Wa	Corr. formula ΔAzim.	Corr. formula ΔElev.	Cause
P2	IA	- IA	-	Azim. encoder zero point
P6	IE	-	+ IE	Elev. encoder zero point
P7	HECE	-	HECE cos e	Gravitational bending
P1	CA	- CA sec e	-	Collimation beam - elev. axis
P3	NPAE	- NPAE tan e	-	Collimation azim. - elev. axis
P4	AW	-AW tan e cos A	+AW sin A	Az. axis (E/W) misalignment
P5	AN	-AN tan e sin A	-AN cos A	Az. axis (N/S) misalignment

Some workers have taken the entirely empirical approach and fitted the observations with multi-parameter functions without regard of their physical reality. While this may yield a superior fit, it would appear to be less satisfactory, because no insight is gained this way in the behaviour of the antenna over time. Using physically realistic terms in the pointing solution enhances the possibility to understand time variable pointing effects such as caused by wind and temperature variations. On the basis of this understanding one can then plan to take corrective action.

5.5.3. Measurement of the optimum focus

In Chapter 4 we have discussed the influence of a defocused situation on the gain and pointing direction of the antenna. We saw that in particular an axial defocus of the prime focus feed or the Cassegrain subreflector leads to a significant loss in antenna gain. While the gain loss with a lateral defocus is less severe, in this case we encounter the strong and asymmetric sidelobe - the coma lobe, along with a pointing error. Thus for an optimum operation of the antenna with maximum gain and low sidelobes, it is important that the feed or Cassegrain secondary reflector be positioned in the true focal point of the primary reflector. We recall that in a Cassegrain antenna, because of the magnification effect, the feed position in the secondary focus is less sensitive (see Ch. 4.3.3). For the reasons described above in the pointing discussion, it will be necessary to regularly check for the optimum focus position, both in the axial and lateral direction. This can be done with relative ease using a celestial radio source of angular size smaller than the HPBW of the antenna at the frequency of operation. For the determination of the axial focus we exploit the rather fast variation of the antenna gain with axial feed position as discussed in Chapter 4.3.1. We saw there (Eqs. (4.22-23) that for relatively small defocus the gain varies as the square of the focal displacement. Our method then consists of a gain measurement at minimally three, preferably five axial positions of the feed, spaced by about 0.2 wavelength. This will lead to a gain loss of about 25 percent at the extreme points. Fitting the measured points with a quadratic curve and finding its maximum will deliver the position of the true focus of the antenna. An example of such a measurement is presented in Fig. 5.13. In this figure the "focus curve" is shown for three cases, which show the influence of the angular size of the source. As explained in Ch 5.3.2 the gain decrease will be smaller in case the source covers a sizeable part of the main beam. In the extreme case where the source is much larger than the beam, as in the example of the Moon in the figure, the gain decrease is very slow, because the power, removed from the main beam and reappearing in the sidelobes is still received in this case. From the measurements it appears that the focus position found with an very large source is slightly different from that when using a point source. This is probably a result of a somewhat asymmetrical sidelobe structure of the defocused beam. The antenna used for these measurement was already at the edge of its performance in terms of surface quality. See also [Mat.5.5].

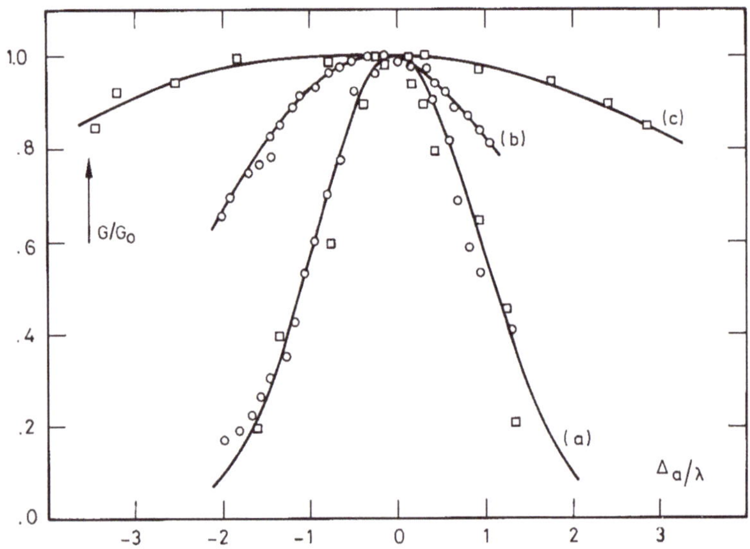

Fig. 5.13. Measured relative antenna gain as function of axial defocus in units of wave length. Curve (a) was obtained on a point source, (b) on an extended source of size twice the beamwidth and (c) on the Moon, 15 times the beamwidth.

Checking the lateral focus can be done in a similar way by moving the feed in lateral direction. This must be done in two orthogonal planes to fully derive the off-axis position. Because the gain loss is much smaller for a certain feed offset than with axial defocus, this measurement is harder to perform accurately. Of course, the weak dependence of the gain on defocus renders the focusing less critical. If a strong source is used it is advantageous to adjust the feed to a position where the level of the first sidelobe is equal in all four directions of defocus. As we saw in Chapter 4, the fast rise of the coma lobe with lateral defocus gives a good impression of the situation.

We have already discussed above the effects of structural deformation as function of elevation angle, which make a focus correction necessary. For a constant load, like gravity, this will be a constant correction over time. Most antennas exhibit in addition a noticeable change in focal length as function of the temperature of the telescope structure. Hence a diurnal variation will be observed which must be corrected. If the thermal time constant of the structural sections is not too small, the relation between focal length and ambient temperature (or the temperature of a major structural part) will be rather constant over time and a fixed correction curve for the focus as function of temperature can be incorporated in the control algorithms. As an example of such corrections the focus data of the ALMA prototype antennas (see Mangum et al, 2006) are summarised in Fig. 5.14. The complete set of focus data could be fitted very well by a combination of a Sin(elevation) function, related to the gravitational bending of the structure, and a linear term with temperature change.

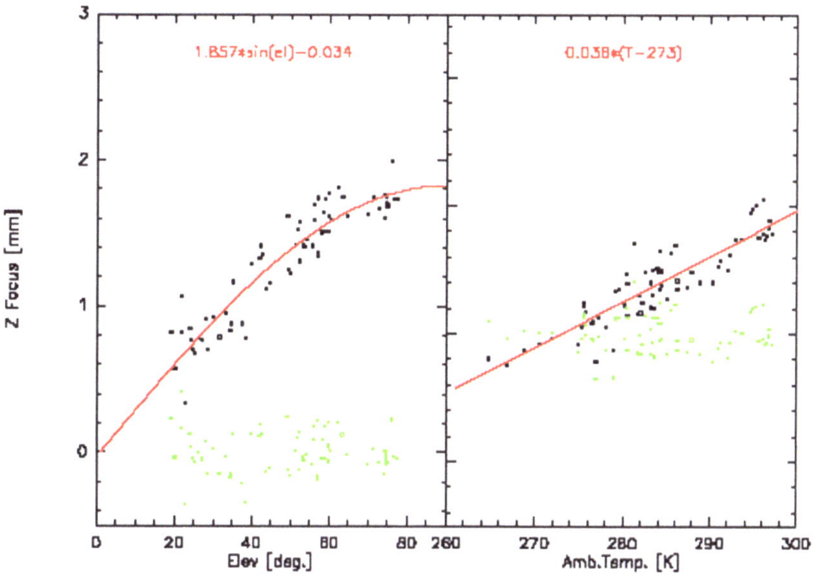

Fig. 5.14. Axial focus curves of the ALMA prototype antenna, separated in a sine term as function of elevation angle, representing gravitational deformation and a linear term with ambient temperature. The functional relationships are printed in the top of the figure. The green dots are the residuals from the fit, about 0.4 mm peak to peak. (After Mangum et al., 2005).

■ 5.6. Pointing and surface error calculation from Finite Element Analysis (FEA)

In foregoing sections we have described the methods and formulas to establish the pointing correction and rms reflector surface error from measurements on radio sources. In the design of a new antenna these errors under operational conditions must be calculated in the structural analysis (finite element analysis, FEA) and the predicted values must remain within the specified bounds. Sometimes confusion has arisen between the structural engineer and the user on the correct method of using the FEA results to predict the operational errors with which the user will be faced. We shall now describe in general terms the method to arrive at useful operational parameters from FEA calculations.

The ideal antenna provides an infinitely stiff, perfectly paraboloidal reflector attached to a perfect mount of infinite stiffness and perpendicularly aligned axes. This provides an accurately known collection area and pointing in the direction of the antenna encoder read-out. In real life, no structure is perfectly dimensioned, nor infinitely stiff and hence there will be mechanical imperfections in fabrication and assembly, as well as structural deformations caused by gravity, temperature gradients and wind forces. These result in a change of pointing direction and a deviation of the reflector surface from the prescribed paraboloid. It are these errors under operational situations which are of main concern to the user and form the main specification.

The errors are basically caused by the path length variations imposed on the incoming wave-front on its way to the focal point by the motions of the points on the reflector (both main reflector and secondary). An appropriate "summing" of the wave-fronts errors for all points of the surface will provide the geometry of the reflector (shape and axis direction) from which the residual surface errors, delay errors and pointing errors can be derived.

It should be noted that a constant path length variation for all points in the incoming wave-front does not lead to any decrease in performance; the antenna is still perfect and well-focused as far as the waves are concerned. In effect, the wave-front naturally "chooses" the most effective path length distribution to concentrate the maximum amount of energy in the "best-fit" focal point, which normally will be different from the original focus position.

Thus, generally we are allowed to consider the structural deviations of the reflector with respect to the "best-fit paraboloid" in the error analysis. For any reasonable structure the residual deformations will be small compared to the size of the deforming component. This allows us to use the wave-front method, where we assume the waves to undergo only path length changes on their original paths. This approach is well suited to the situation with large structural elements and small deformations, both in units of the wavelength.

The best-fit paraboloid will be shifted in space, both in translation and rotation, with respect to the original reflector and will have a different focal length. This leads to the following error components:

1. a pointing error, caused by the shift of the reflector axis direction, together with any lateral movements and rotation of the subreflector and shift of the receiver position.

2. a delay error from the axial movements of the reflector, subreflector and receiver and from the change in the focal length of the primary.

3. a residual surface error, being the difference between the real surface coordinate and that of the best-fit paraboloid at the corresponding point.

The basic specification of the antenna sets maximum values to these errors. To exploit the best-fit procedure, it will be necessary to move the subreflector to a new position which minimises the residual surface errors; this implies a movement in all three coordinates, two for pointing, one for focus. In practice this is achieved by moving the subreflector to a position where its imaginary focus coincides with the focus of the paraboloid which "best" fits the shape of the deformed primary.

With regard to the use of the best-fit paraboloid, the following should be considered. It is allowed to use this procedure for all repeatable, time invariable effects, gravity being the major contributor. For time variable influences, as temperature changes and wind forces, each condition must be considered separately for pointing error and surface error.

Temperature varies relatively slowly in time and most of its effects on pointing and surface error will be minimised by the regular pointing and focus "calibration". The residuals after this calibration should be used in the error tabulation.

Any error caused by short term influences, wind in particular, will not be "corrected" by calibration and must be accounted for fully in the error analysis. Thus for wind deformation the best-fit approach cannot be used for the residual reflector surface error, because normally we cannot measure or calibrate the new parameters of the best-fit reflector. Only a fast (time constant of order seconds), closed loop measurement and adjustment of the reflector surface could achieve this. Such "active surface" solutions are now being tried, but have not reached the speed needed for correction of wind induced deformations. Thus the residual surface errors should be calculated without allowing for the possibility of fitting a paraboloid with a different focal length from that of the primary reflector without wind loads.

For pointing the situation is similar. As mentioned above, the wave-front will seek the best-fit paraboloid and concentrate its energy to its focal point. The pointing error in wind will now be the difference between the direction of this "moving" focus and that which we used as starting point. The full value of this error must be taken into the error tabulation.

We can now summarise the situation with the following "**recipe**" for the error analysis based on the FEA results for all operational conditions.

1. The FEM delivers the three-dimensional coordinates of the reflector, the subreflector/quadripod and the receiver as function of the loading condition.

2. From this we compute the location and orientation in space of the best-fit paraboloid and of the subreflector. The paraboloid is characterised by 6 parameters: x,y,z-coordinates of the reflector vertex, rotation about the x and y axis and focal length. The subreflector is normally assumed rigid and is characterised by the three translational and two rotational coordinates.

3. These values form the input to the "pointing formulae" provided in Chapter 5.5.

4. The reflector parameters also allow the computation of any delay change.

5. We also can calculate the value of the reflector surface deviation in each computed node as the difference between the real coordinates and those of the corresponding point on the best-fit paraboloid. From this vector we calculate the "axial" component as explained in Chapter 4.6. The rms-value of these is the "surface error". Its effect on the aperture efficiency of the antenna has been discussed in Ch. 4.6 and Ch. 5.4.

5.7. Antenna gain and radio source flux calibration

In the foregoing sections we have often referred to celestial radio sources to be used as "test transmitters" for the measurement of the parameters of large reflector antennas. We shall now have a somewhat closer look at the requirements for such sources to be useful as calibrator sources. We must thereby discern two situations:

i) measurement of absolute antenna parameters - gain, beamwidth, pointing direction

ii) measurement of relative characteristics - beam shape, sidelobe level, position dependent gain variation.

In general we will be interested to determine the antenna parameters as function of frequency and possibly also for varying pointing position. The major advantages and disadvantages of cosmic radio sources for the calibration of antennas can be listed as follows.

Advantages:

- always in the farfield, hence no distance dependent corrections
- diurnal path along the sky provides a range of pointing directions
- normally of small angular size which minimises size corrections
- broadband "continuum" radiation, useful over large frequency range
- normally (but not always) not, i.e. randomly, polarised.

Disadvantages:

- too small an intensity to provide sufficient signal to noise ratio
- irregular or unknown brightness distribution, or too large angular size
- uncertainty in spectrum, i.e. intensity, as function of frequency
- inaccurate knowledge of absolute intensity and/or celestial position
- intensity variations over time (normally very slow, but real).

It will be clear from this list that the use of cosmic sources for antenna calibration is a mixed blessing. It is, however, often the only way to obtain information on the antenna parameters, in particular for antennas with large diameters. Thus, while their use has been a natural approach by radio astronomers, communication engineers and ground station managers have adopted radio sources as test transmitters for their antennas as well.

5.7.1. Determining the absolute gain of the antenna

Perhaps the most important antenna parameter is its gain, or aperture efficiency, at a particular frequency of operation. Only with a known antenna gain will it be possible to obtain the true flux density or brightness of the astronomical source from the measured antenna temperature.

Satellite communication ground stations have strict gain specifications, which must be demonstrated upon delivery to satisfy the defined minimum signal to noise ratio required for a successful commercial operation. There is thus a need for a celestial source of small angular size (a "point source") with precisely known flux density of sufficient intensity, preferably over a broad frequency band and constant in time. Because of the importance of such a calibration source, we summarise here the efforts which have been extended in establishing the absolute spectra of a small number of strong radio sources.

Obviously, a measurement of the *absolute* flux density of a source requires an antenna with an accurately known gain. In addition, the receiver system must be capable of accurately determining the antenna temperature at its input terminals. This is normally done by comparison of the received signal with the power from a matched load at certain, known physical temperature connected to the input terminals of the receiver. This latter activity is in itself far from a routine matter, but it falls outside the scope of this book. Good descriptions of the methods of receiver calibration can be found in Kraus (1966), Findlay (1966) and Ulich and Haas (1976).

As we have seen earlier, a solely theoretical calculation of the gain of a reflector antenna is very difficult and normally not achievable to better than ~10 percent. Thus the use of reflectors to establish absolute flux density will be restricted to such with a size, small enough to be calibrated on an earth-bound test range, the length of which must satisfy the farfield condition Eq. (1.1).

Because of its simple geometrical configuration, an electro-magnetic *horn* is however amenable to an accurate theoretical gain calculation (Schelkunoff, 1943). Several of such horns have been constructed with the aim to establish the absolute flux density of the strongest celestial sources. We mentioned already in Chapter 1 the "Little Big Horn", which had a physical aperture of 20 m^2, operating at 1400 MHz (Findlay et al., 1965). This horn was constructed with the specific aim of making an absolute flux density measurement of the strongest radio source of small angular size, Cassiopeia A (Cas A). Other groups constructed horns for this purpose and made measurements at a number of frequencies between 2 and 16 GHz. Some details on the methods employed are described by Findlay (1966). These experiments have shown that with careful fabrication and experiment horn measurements can achieve an accuracy of about 2 percent.

At frequencies well below 1 GHz it is feasible to use relative small dipole arrays over a ground-plane. The gain of such arrays can also be calculated theoretically with 1-2 percent accuracy. Several measurements of the strong sources Cas A, Cyg A and Tau A have been made in this way with 2-3 percent accuracy (Parker, 1968; Wyllie, 1969).

An original way of calibrating relatively small parabolic reflectors has been devised by the Troitskii and Tseitlin (1962) group at the Scientific-Research Radiophysical Institute, Gorki (USSR). It is known as the "*artificial moon method*" and works as follows (for a summary description see Findlay, 1966). On a hilltop at about 500 m from the antenna under test at an elevation angle of 20° a "black disc", covered with absorbing material, is placed in the beam of the antenna. Measurements were made at wavelengths between 1.5 and 10 cm with 4 and 1.5 m diameter dishes. Thus the black disc, or artificial moon, is in the farfield of the antennas, while the disc has diameters between 1 and 4.6 m, depending on the wavelength. The disc subtends an angle at the antenna of about one third of the HPBW. The mechanisms are built in such a way that the disc can be moved in and out of the beam without altering the surroundings of the disc in any way. What is being measured at the antenna terminals is the difference in signal from the disc and the sky behind the disc. Because of the relatively small wavelength, the latter is small. It is the 3 K cosmic background radiation plus a small Galactic component (relatively well known from sky surveys) and the atmosphere behind the disk. The disk itself is near 300 K, hence a small uncertainly in the weak background will not cause a grave error. Applying now the formulas of Ch. 5.3.2, we find that the measured difference in antenna temperature is given by

$$\Delta T = \eta_R \, (T_{\rm disc} - T_{\rm sky}) \, \frac{1}{\Omega_A} \int_{\Omega_{\rm disc}} g(\theta, \phi) \, \psi(\theta, \phi) \, d\Omega \; = \; (T_{\rm disc} - T_{\rm sky}) \, \frac{A}{\lambda^2} \, \Omega_{\Sigma{\rm disc}} \tag{5.48}$$

where all variables have their usual meaning and $\Omega_{\Sigma\mathrm{disc}}$ is the beam weighted source solid angle of Eq. (5.25), calculated from the known antenna beamwidth and the disc angular extent as seen from the antenna. Thus from the measured antenna temperature difference ΔT, the absorption area A of the antenna can be found. In the original experiment, the accuracy of the method is quoted to be between 5 and 10 percent. Improvements have been made over the years and the method is now said to be accurate to 3-4 percent (Ivanov and Sharova, 2002).

5.7.2. Extraterrestrial sources as test transmitters

The methods just mentioned have been used to establish the absolute spectrum of a few strong radio sources, the strongest and best studied being Cassiopeia A (Cas A). Unfortunately Cas A is not at all ideal for this purpose. Its angular size of about 4 arcminutes requires source size corrections for beamwidhts of the same order of magnitude, which are common now with large telescopes at short wavelengths. Also, Cas A being a recent supernova remnant, exhibits a secular decrease in flux density of about 1 percent per annum, which moreover is dependent on frequency. However, because of the importance of the object, it has been extensively and carefully studied, resulting in a well-known spectrum between 20 MHz and 30 GHz and an accurate functional form for the secular decrease. In the paper by Baars et al. (1977) its spectrum, along with those of Cyg A and Tau A, was established with a mean error of 2 percent between 300 MHz and 30 GHz from about 50 absolute measurements. This spectrum forms the basis for the flux density scale used in radio astronomy. These strong sources have also been used extensively for the gain calibration of satellite communication ground stations. A so-called G/T-determination (the ratio of gain to system temperature, which defines the performance of a communication channel) of ground stations is routinely done with a measurement on Cas A, making use of the spectrum and absolute scale, established by the radio astronomy community.

We present now the data on the spectrum of Cas A. Later in this section, we shall turn to the subject of source flux density calibration by introducing a set of secondary standard sources, that do not exhibit the disadvantages listed above and are sufficiently strong for the highly sensitive observing systems, including interferometer arrays. The spectrum of Cas A, as determined by Baars et al. (1977) can be written as

$$\log S\,[\mathrm{Jy}] = a + b \log \nu\,[\mathrm{MHz}] + c \log^2 \nu\,[\mathrm{MHz}]\ , \text{ with}$$

$a = 5.625 \pm 0.021$, $b = -0.634 \pm 0.015$, $c = -0.023 \pm 0.001$ for $22 < \nu < 300$ MHz,

$a = 5.880 \pm 0.025$, $b = -0.792 \pm 0.007$, $c = 0$ for 300 MHz $< \nu <$ 31 GHz at the epoch 1965.0. The secular decrease as function of frequency is given by

$$d(\nu)\,[\%\text{ per year}] = 0.97\,(\pm 0.04) - 0.30\,(\pm 0.03) \log \nu\,[\mathrm{GHz}].$$

Thus for epoch 2010, the "power-linear" spectrum for the frequency range 0.3 - 31 GHz would be

$$\log S\,[\mathrm{Jy}] = 3.255 - 0.701 \log \nu\,[\mathrm{MHz}]$$

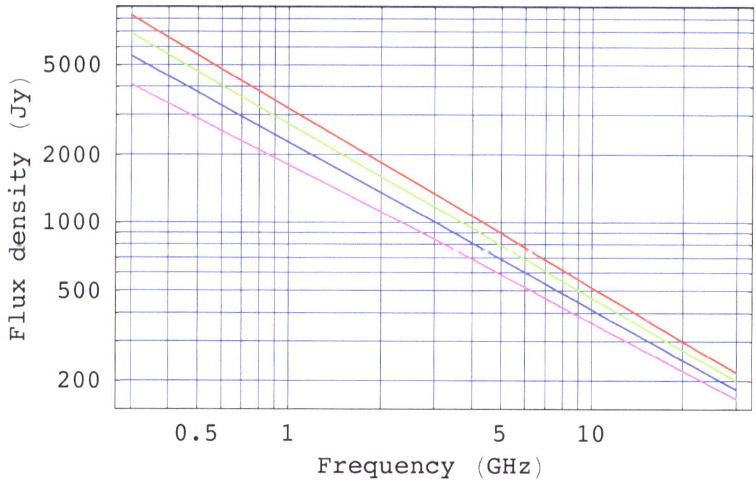

Fig. 5.15. The flux density of Cassiopeia A as function of frequency for the years 1965, step 15, 2010, top to bottom. The relative secular decrease is smaller at higher frequencies.

From Fig. 5.15 [Mat.5.6] we see that the flux density has become less than 200 Jy for frequencies above 30 GHz. For the calibration of millimeter wave antennas a stronger source would be of advantage. Moreover, the angular size of Cas A of about 4 arcminutes results in considerable and uncertain size corrections with the large telescope and ground stations of about 30 m diameter, which have a HPBW ≈ 1.5 arcminute at 30 GHz. Because of the steep spectrum of Cas A, the situation becomes even more critical at frequencies of 100 GHz and higher.

The **planets** offer themselves as calibrators for the millimeter and submillimeter frequency range. They radiate essentially as black bodies in this frequency range, which means that their brightness temperature is constant and hence the flux density will increase proportional to the square of the frequency. While they do not have a fixed celestial position and exhibit a varying angular size because of their orbits around the Sun, their ephemerides are well known and available in tabular form. The planetary temperatures have been the subject of many studies and they often show interesting, but for our purpose complicating, deviations from pure black-body radiation. An important step was taken by Ulich et al.(1980), who made an accurate determination of the brightness temperatures of Venus, Jupiter and Saturn at 3.5 mm wavelength with an accuracy of about 3 percent. The result is shown in Table 5.5. At shorter wavelengths these planets behave essentially as black bodies, although the size and varying inclination of the rings of Saturn can pose a problem for narrow beam widths of less than a minute of arc (Mangum, 1993). Mars behaves very closely as a black body over the entire mm- and submm-wavelength range. According to Ulich (1981) its brightness temperature varies proportional to $\sqrt{R_0 / R}$, where R is the Martian heliocentric distance and R_0 = 1.524 AU (astronomical unit), the mean heliocentric distance. Its brightness temperature at R_0 is added to Table 5.5.

A limited amount of work in this area has been done since then, particularly at BIMA (Welch et al, 1996), where some absolute planetary brightness temperatures at 11 and 3 mm wavelength to an accuracy of 1-2 percent were established (Gibson et al., 2005; Gibson and Welch, pers. comm., Sep. 2005). Extension of the flux density scale to short mm-wavelengths is in need of improvement. New telescopes for this frequency range, in particular ALMA, will certainly increase activity in this field.

Table 5.5. Brightness temperatures at 3.5 mm wavelength

Object	Brightness Temperature (K)	Error (1σ)
Sun	7914	192
Venus	357.5	13.1
Jupiter	179.4	4.7
Saturn	153.4	4.8
Mars	206.8	5.8

For the radio astronomer, the knowledge of the absolute gain of the antenna is not of prime importance. Rather his interest is in knowing the absolute intensity of the observed object in order to connect his observation to intrinsic source parameters. This can be achieved most easily by comparing the received signal with that from a reference source with absolutely known flux density. It was for those reasons that the strongest sources were measured with absolutely calibrated antennas. As we saw already, those few strong sources are not ideal for routine comparison work. Thus considerable effort has been devoted to establishing a set of *secondary standards* which are more suited for daily use as flux density calibrators. The spectra (the intensity as function of frequency) of these sources have been determined by careful measurements of the ratio of their flux density to that of one of the major standards

(Baars et al., 1977). The secondary standards have been chosen to be of intermediate strength in order to avoid problems with alinearity in the receiver system and to be distributed over the celestial sphere, so normally a standard will be available within a reasonable angular distance from the source being observed. In addition one has given preference to sources of very small angular extent (so-called point sources) which avoids the need for size corrections to the measured flux density. This has the additional advantage that they can also be used for the calibration of interferometers and synthesis telescopes, which can have an angular resolution of better than 1 arcsecond. Finally, the positions of these sources have also been determined very accurately and hence they can be used as calibrators for antenna pointing and interferometer baseline calibration. Thompson et al. (2001) discuss this in detail.

■ 5.8. The Mathematica routines

```
Mat .5 .1 - ext. source correction factors;
kd = (x / 1.2) ^ 2 / (1 - Exp[- (x / 1.2) ^ 2]) ; kg = 1 + x ^ 2;
θg = Sqrt[1 + x ^ 2]; θd = Sqrt[1 + (Log[2] / 2) x ^ 2];
Plot[Evaluate[{kg, kd, θg, θd}, {x, 0, 1},
   PlotRange → All, GridLines → Automatic, Frame → True,
   FrameLabel → {"source size/beam width", "Corr.Factor"},
   PlotStyle → {{RGBColor[1, 0, 0]}, {RGBColor[0, 1, 0]},
      {Dashing[{.05, .05}], RGBColor[0, 0, 1]},
      {Dashing[{.05, .05}], RGBColor[1, 0, 1]}}]]
```

```
Mat .5 .2 -
   convolution with straight edge of gauss and lambda beams;
θ = 4; x =.; u =.;
con = Integrate[
    Exp[- ((x - u) / (.6006 θ)) ^ 2] UnitStep[u], {u, -50, 50}];
cov = Integrate[ 2 BesselJ[1, (x - u)] / (x - u) UnitStep[u],
    {u, -50, 50}];
din = D[con, x];
div = D[cov, x];
cv =
  Plot[{con, cov}, {x, -10, 10}, DisplayFunction → Identity,
   GridLines -> Automatic, Frame -> True,
   PlotStyle → {{RGBColor[1, 0, 0]}, {RGBColor[0, 0, 1]}}]
di = Plot[{din, div}, {x, -10, 10}, PlotRange → All,
   GridLines -> Automatic, Frame -> True,
   PlotStyle → {{RGBColor[1, 0, 0]}, {RGBColor[0, 0, 1]}},
   DisplayFunction → Identity]
Show[GraphicsArray[{cv, di}]]
```

Comparison of gaussian and lambda function representation of the antenna beam [Mat.5.3].

[Mat.5.3] provides a comparison between the gaussian beam approximation and the "Lambda-function" result of the integration over the aperture distribution (see Eq. (4.12) with a realistic quadratic shape and taper of -12 dB at the edge ($\tau = 0.25$). The half-power width of the two functions has been forced to be equal at x = 1, which causes the factor $u = 1.82\,x$ in the Bessel function representation. The plots in Fig. 5.16 with linear and logarithmic intensity scale illustrate the differences. The beams are essentially coincident down to a level of 0.2 in power (-7 dB) at an angular coordinate of 1.5 times the HPBW. Beyond that point the gaussian approximation overestimates the beam intensity, but by then the beam is at a level of -20 dB. A numerical integration of the Lambda-function representation up to its first null at x = 2.52 yields a beam solid angle which is 2.8% smaller than the value of Eq. (5.28) for the gaussian approximation. Note that the gaussian representation does not show discrete sidelobes. This is obvious, because the Fourier Transformation (see Ch. 3.7) of a gaussian function is again a gaussian, which descends continuously.

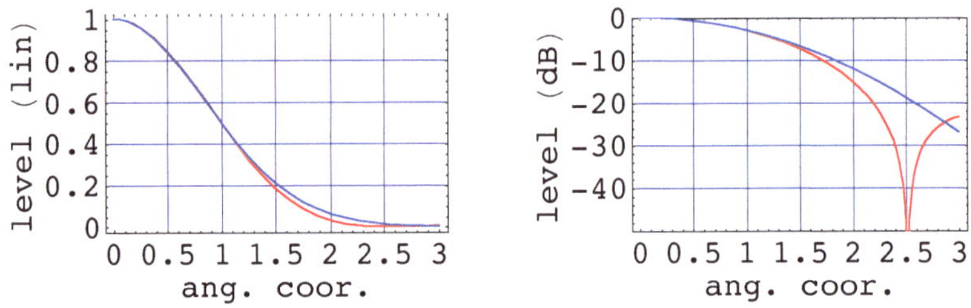

Fig. 5.16. The main beam with normalised half-power width for a gaussian (blue) and tapered Lambda-function (red) approximation.The beams are nearly identical down to a level of-7 dB.

```
Mat .5 .3 - Gaussian and Lambda beam;
τ = 0.25; u = 1.82 x;
gl = (2/(1 + τ) (2 τ BesselJ[1, u] / u + 4 (1 - τ) BesselJ[2, u] / u^2))^2 ;
gg = Exp[- (x ^ 2 / (0.6006 2) ^ 2)];
pi =
 Plot[Evaluate[{gl, gg}, {x, 0, 3}, DisplayFunction → Identity,
   GridLines -> Automatic, Frame -> True,
   FrameLabel -> {"ang. coor.", "level (lin)"},
   PlotStyle -> {{RGBColor[1, 0, 0]}, {RGBColor[0, 0, 1]}}]]
pl = Plot[Evaluate[10 Log[10, {gl, gg}], {x, 0, 3},
   PlotRange -> {-50, 0}, DisplayFunction → Identity,
   GridLines -> Automatic, Frame -> True,
   FrameLabel -> {"ang. coor.", "level (dB)"},
   PlotStyle -> {{RGBColor[1, 0, 0]}, {RGBColor[0, 0, 1]}}]]
Show[GraphicsArray[{pi, pl}]]
```

```
Mat .5 .4 a - data preparation and fit;
datl = {3.5, 2.75, 2.1, 1.3, 1.13, .87}; di = 1 / %^2;
datf = {.53, .5, .45, .27, .21, .1}; dl = Log[%];
dat = {{di[[1]], dl[[1]]}, {di[[2]], dl[[2]]},
    {di[[3]], dl[[3]]}, {di[[4]], dl[[4]]},
    {di[[5]], dl[[5]]}, {di[[6]], dl[[6]]}};
fd = Fit[dat, {1, x}, x]
```

Out[100]= -0.509658 - 1.35212 x

```
Mat .5 .4 b - plot of error calculation;
ε = Sqrt[1.35212] / (4 π)
ηA0 = Exp[-.51]
lpd = ListPlot[dat, PlotStyle → PointSize[0.015]]
pfd = Plot[fd, {x, 0, 1.5}]
Show[{pfd, lpd}, Frame → True,
 GridLines → Automatic, FrameLabel → {"1/λ²", "Log ηA"}]
```

Out[108]= 0.0925332

Out[109]= 0.600496

```
Mat .5 .6 - Radio spectrum of Cassiopeia A;
<< Graphics`
Sν in Jy; ν in GHz;
ν =.; t = AD - 1965; d = t (0.97 - 0.3 Log[10, ν]) / 100;
Sν = (1 - d) * 10^((5.88 - 0.792 Log[10, 1000 ν]));
fl = Table[Sν, {ν, 1, 31, 10}, {AD, 1965, 2010, 15}]
data = {{1, fl[[1, 4]]}, {11, fl[[2, 4]]}, {21, fl[[3, 4]]}};
Fit[Log[10, data], {1, x}, x]
LogLogPlot[Evaluate[Table[Sν, {AD, 1965, 2010, 15}],
  {ν, .3, 30}, Frame → True, GridLines -> Automatic,
  FrameLabel → {"Frequency (GHz)", "Flux density (Jy)"},
  PlotStyle → {{RGBColor[1, 0, 0]}, {RGBColor[0, 1, 0]},
    {RGBColor[0, 0, 1]}, {RGBColor[1, 0, 1]}}]]
```

```
Mat .5 .5 - simulation of pointing measurement;
data = {{-2, 7.4}, {-1, 8.}, {0, 8}, {1, 7.1}, {2, 5}};
ld = ListPlot[data, PlotStyle → PointSize[0.02]]
fd = Fit[data, {1, x, x^2}, x]
FindMaximum[fd, {x, 0}]
pf = Plot[fd, {x, -3, 3}, DisplayFunction → Identity]
Show[ld, pf]
```

```
Out[3]=  8. - 0.57 x - 0.45 x^2
Out[4]=  {8.1805, {x → -0.633333}}
```

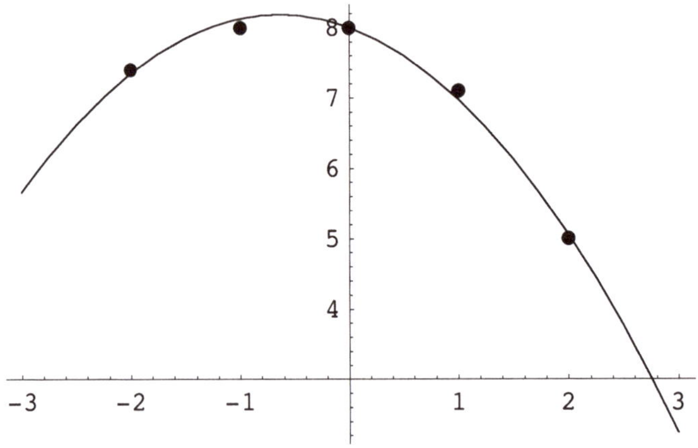

This is a routine which simulates the data from a pointing measurement consisting of a five-point raster scan through the source with data taken in the nominal focus position and 1 and 2 units shift on either side, respectively. A quadratic function forms a very good fit and the maximum is found at a position of -0.633.

References

Abramowitz, M. and I.A. Stegun, *Handbook of Mathematical Functions*, Nat. Bureau of Standards, 1964.

Baars, J.W.M., The Measurement of large Antennas with cosmic Radio Sources, *IEEE Trans. Antennas Propagat.* **AP-21**, 461-474, 1973 [Gold, 171].

Baars, J.W.M., J.F. van der Brugge, J.L. Casse, J.P. Hamaker, L.H. Sondaar, J.J. Visser and K.J. Wellington, The Synthesis Radio Telescope at Westerbork. *Proc IEEE* **61**, 1258-1266, 1973 [Gold, 116].

Baars, J.W.M., R. Genzel, I.I.K. Pauliny-Toth and A. Witzel, The Absolute Spectrum of Cas A; An accurate flux density scale and a set of secondary calibrators. *Astron. Astrophys.* **61**, 99-106, 1977 [Gold, 401].

Bracewell, R., *The Fourier Transform and its Applications*, New York, McGraw-Hill, 1965.

Findlay, J.W., H. Hvatum and W. B., Waltman, An absolute flux density measurement of Cassiopeia A at 1440 MHz. *Asrophys. J.* **141**, 873-884, 1965.

Findlay, J.W., *Absolute Intensity Calibrations in Radio Astronomy*, Ann. Rev. Astron. Astrophys. **4**, 77-94, 1966.

Gibson, J., W.J. Welch and Imke de Pater, Accurate jovian radio flux density measurements show ammonia to be subsaturated in the upper troposphere, *Icarus* **173**, 439-446, 2005.

Gordon, M.A., J.W.M. Baars and J. Cocke, Observations of radio lines from unresolved sources: telescope coupling, Doppler effects and cosmological corrections, *Astron. Astrophys.* **264**, 337-344, 1992.

Greve, A., C. Kramer and W. Wild, The beam pattern of the IRAM 30-m telescope, *Astron. Astrophys. Sup.* **133**, 271-284, 1998.

Ivanov, V.P. and O.I. Sharova, Reference spectra in the "artificial-moon" scale and cosmological studies of radio sources, *Radiophys. and Quantum Electr.* **45**, 91-101, 2002.

Kraus, J.D., *Radio Astronomy,* New York, McGraw-Hill, 1966.

Kutner, M.L. and B.L. Ulich, Recommendations for calibration of millimeter-wavelength spectra line data, *Astrophys. J.* **250**, 341-348, 1981 [Gold, 423].

Mangum, J.G., Main-Beam Efficiency Measurements of the Caltech Submillimeter Observatory, *Publ. Astron. Soc. Pacific* **105**, 117-122, 1993.

Mangum, J.G., J.W.M. Baars, A. Greve, R. Lucas, R.C. Snel, P. Wallace and M. Holdaway, Evaluation of the ALMA Prototype Antennas, *Proc. Astron. Soc. Pacific* **118,** 1257-1301, 2006.

Nyquist, H., Thermal agitation of electric charge in conductors, *Phys. Rev.* **32**, 110-113, 1928.

Parker, E.A., Precise measurement of the flux densities of the radio sources Cas A and Cyg A at metre wavelengths, *Monthly Notices Roy. Astron. Soc.* **138**, 407-422, 1968

Ruze, J., The effect of aperture errors on the antenna radiation pattern, *Suppl. al Nouvo Cimento* **9**, 364-380, 1952.

Ruze, J., Antenna tolerance theory, *Proc. IEEE* **54**, 633-640, 1966 [Gold, 185].

Schelkunoff, S.A., *Electromagnetic Waves*, 363-365, New York, van Nostrand, 1943.

Smart, W.M., *Spherical Astronomy,* Cambridge, University Press, 1962.

Stumpff, K., *Geographische Ortsbestimmungen*, 35, Berlin, VEB Deutscher Verlag de Wissenschaften, 1955.

Stumpff, P., *Astronomische Pointingtheorie für Radioteleskope*, Kleinheubacher Berichte **15**, 431-437, 1972.

Troitskii, V.S. and N.M. Tseitlin, Application of Radio-astronomical Method for Calibrating small Antenna Systems at centimeter Wavelengths (in Russian), *Radiofysika* **5**, 623-628, 1962.

Thompson, A.R., J.M. Moran and G.W. Swenson, *Interferometry and Synthesis in Radio Astronomy*, 2nd Ed., New York, John Wiley, 2001.

Ulich, B.L., Millimeter-wavelength continuum calibration sources, *Astron.J.* **86**, 1619-1626, 1981.

Ulich, B.L. and R.W. Haas, Absolute Calibration of Millimeter-wavelength spectral Lines, *Astrophys. J. Suppl.* **30**, 247-258, 1976.

Ulich, B.L., J.H. Davis, P.J. Rhodes and J.M. Hollis, Absolute Brightness Temperature Measurements at 3.5-mm Wavelength, *IEEE Trans. Antennas Propagat.* **AP-28**, 367-377, 1980 [Gold, 413].

Wallace, P.T., Pointing and tracking software for the Gemini 8-m telescopes, *Proc. SPIE* **2871**, 1020, 1997.

Wallace, P.T., A rigorous algorithm for telescope pointing, *Proc. SPIE* **4848**, 125, 2002.

Welch, W. J. et 36 alii, The Berkeley-Illinois-Maryland-Association Millimeter Array, *Publ. Astron. Soc. Pacific* **108**, 93-103, 1996.

Wyllie, D.V., Absolute flux density scale at 408 MHz, *Monthly Notices Roy. Astron. Soc.* **142**, 229-240, 1969.

6. Miscellaneous subjects

6.1. Holographic measurement of reflector surface contour

6.1.1. Introduction

Large reflector antennas, as those used in radio astronomy and deep-space communication, generally are composed of a set of surface panels, supported on three or more adjustable points by a support structure, called the backup structure. After assembly of the reflector it is necessary to accurately locate the panels onto the prescribed surface in order to obtain the maximum antenna gain. The fact that some antennas have a "shaped" contour is irrelevant for the purpose of our discussion. We are concerned with describing a method which allows us to derive the position of the individual panels in space and compute the necessary adjustments of their support points to obtain a continuous surface of a certain prescribed shape.

In Chapter 4.6.1 we have discussed the influence of random errors in the reflector contour on the antenna gain. We found that an error of $\lambda/40$ is required to limit the gain loss to 10 percent, while for an error of $\lambda/16$ the gain is decreased to about half of the maximum achievable. Radio telescopes are often designed to meet the latter surface error for the shortest operational wavelength. For submillimeter telescopes, operating at 350 μm wavelength, this means a reflector error of 20 - 25 μm. Achieving this is a great challenge both in the design and construction as well as the measurement and setting of the panels.

The first large radio telescopes had a diameter of about 25 m and operated at wavelengths longer than 10 cm. Thus a surface precision of several millimeters provided excellent performance. A measurement accuracy of this order of magnitude is readily achievable with a classic "theodolite and tape" method. Using the best theodolites (e.g. the Wild T3), accuracies of the order of 100 μm have been achieved on reflectors of a size up to 30 m (Greve, 1986). However, the development of the technology of large and simultaneously highly accurate antennas has been a very active field over the last 30 years, whereby the application of the design principle of homology (von Hoerner, 1967) has enabled the construction of, for instance, a 100 m diameter radio telescope with a surface accuracy of 0.5 mm (Hachenberg et al., 1973; Godwin et al., 1986), a 30 m millimeter telescope with 75 μm accuracy (Baars et al., 1987, 1994) and 10-12 m diameter submillimeter telescopes with an rms surface error of less than 20 μm (e.g. Woody et al., 1998, Baars et al., 1999, Smith et al., 2001). The setting of the reflector panels at such accuracy has required the development of measuring methods of hitherto unsurpassed accuracy. It should be noted that these measurements need to be done "in the field", which in the case of millimeter radio telescopes generally means the hostile environment of a high mountain site.

A number of special measuring methods and devices have been developed (for a review, see Baars, 1983). The most versatile, and by now widely used method is normally called "radio holography". The method makes use of a well-known relationship in antenna theory: the farfield radiation pattern of a reflector antenna is the Fourier Transformation of the field distribution in the aperture plane of the antenna. We have met this relationship already in Chapter 3.7. Remember that this relationship applies to the amplitude/phase distributions, not to the power pattern. Thus, if we can measure the radiation pattern, *in amplitude and phase,* over a sufficiently large angular area, we can derive by Fourier Transformation the amplitude and phase distribution in the antenna aperture plane with a certain spatial resolution. The latter is determined by the angular size of the measured radiation pattern. This method was suggested, but not worked out in any detail, in the appendix of Jennison's pocket book "Radio Astronomy" (1966). The paper by Bennett et al. (1976) presented a sufficiently detailed analysis to draw the attention of radio astronomers. Thus, Scott and Ryle (1977) applied it to the new Cambridge 5 km interferometric array to measure the shape of four of the eight antennas, using a celestial radio point source and the remaining four antennas to provide the reference signal for the measurement of the amplitude and phase pattern.

The use of a natural, celestial signal source is very attractive for two reasons. First the source is definitely in the farfield of the antenna. The **farfield region** of the antenna is defined to start at

$$R_f = 2d^2 / \lambda, \tag{6.1}$$

and can easily reach values of several hundreds of kilometers. Thus an earth-bound transmitter will hardly ever be in the farfield for these applications. Secondly, the celestial radio source traces a daily path across the sky, providing a range of elevation angles over which the data can be collected. This is of great interest for the study of elevation dependent deformations of the antenna, caused by gravity. However, often the intensity of the cosmic source is not sufficient to achieve the required signal to noise ratio for an accurate measurement. Only a few strong sources are available. The situation is more favorable, if there are several large antennas, as in interferometric arrays, where the extra antennas can be used to provide a strong reference signal.

For the IRAM 30 m millimeter telescope on Pico Veleta (Baars et al., 1987) it was decided to use a holographic system at 22 GHz, using the very strong water vapour maser source in the Orion Nebula, which flared to an intensity of several million jansky at the time of the design phase of the telescope. The reference signal was provided by a 1.5 m diameter reflector located in the back of the prime-focus cage of the telescope. A compact dual receiver in the prime focus served both reference and main reflector. Although by the time of the measurement the maser source had weakened, it was still sufficiently strong to enable a measurement of the surface with an accuracy of about 30 μm rms and a setting of the surface to better than 100 μm rms (Morris et al., 1988).

Artificial satellites, radiating a beacon signal at a fixed frequency can also be used as farfield signal source. Extensive use has been made of synchronous communication satellites in the 11 GHz band (e.g. Godwin et al., 1986). These fixed transmitters do not provide the range of elevation angles accessible with cosmic sources. Some satellites, notably the LES (Lincoln Experimental Satellite) 8 and 9, have been used for radio holography of millimeter telescopes (Baars et al., 1999). They provided a signal at the high frequency of 37 GHz and with their geo-synchronous orbit moved over some 60 degrees in elevation angle. Unfortunately, both satellites are no longer available. Radio astronomers would be greatly helped if a satellite would become available with a reliable transmitter of adequate power at a high frequency of about 40 GHz, or preferably near 95 GHz. To avoid interference of regular radio astronomical observations at this frequency, the transmitter should normally be switched off and only be available at internationally coordinated time intervals.

Lacking a sufficiently strong source in the farfield, we have to take recourse to using an earth-bound transmitter. In practice it will be located at a distance of several hundreds of meters to a few kilometers and be at an elevation angle of less than 10 degrees. Clearly, these are in the nearfield of the antenna, requiring significant corrections to the received signals. A detailed treatment of this case is presented here. Successful measurements on short ranges have been reported for the JCMT (Hills et al., 2002) and the ASTE antenna of NAOJ. The ALMA (Mangum et al. 2006) prototype antennas (12 m diameter, surface accuracy specification 20-25 μm) have been measured and set with the aid of a transmitter at a distance of only 315 m, elevation angle 9 degrees, radiating at a wavelength near 3 mm. We illustrate the discussion below with examples from these measurements (Baars et al. 2006).

6.1.2. The mathematical basis of Radio Holography

The basic expression, linking the radiation function $f(x, y, z)$ at a point P in space with the field distribution $F(\xi, \eta)$ over the aperture plane of the antenna has been derived in Chapter 3. There we also concluded that this relation is basically a Fourier Transformation expression. Thus we can write the inverse Fourier Transformation as follows (see also Ch. 3.7)

$$F(\xi, \eta) = \int\int f(u, v) \frac{e^{ikr}}{r} \exp\{-ik(u\xi + v\eta)\} \, du \, dv, \tag{6.2}$$

where (u, v) are the direction cosines towards a point at the distance r from the aperture. In principle, the integration has to be performed over the entire sphere, surrounding the aperture. Thus a knowledge of the complete farfield radiation function, both in amplitude and phase will provide us with the complex field distribution (amplitude and phase) over the aperture of the antenna. In most cases it will be impossible, or in any case impractical, to measure the farfield pattern over the entire sphere. Fortunately, this is not necessary for the purpose of deriving the shape of the reflector in sufficient detail. With Nyquist's sampling theorem in mind it can be seen that a measurement of the beam pattern over an angle n times the beamwidth (λ/d) will provide information of the aperture distribution on a spatial scale of d/n, where d is the aperture diameter. The same scale will be achieved in our knowledge of the reflector panel position. Thus as long as we have several scale-lengths over the linear extent of the panel, its overall position with respect to the prescribed contour can be determined. Of course, panel deformations on a smaller scale will not be detected, but these would not be amenable for adjustment with the typically 3-5 support points per panel. Most antennas have between 5 and 15 rings of panels and a value of n ~ 100 will be fully adequate.

If the signal source is in the farfield, the practical realisation of the method is rather straightforward, because few corrections have to be applied. However, the signal-to-noise ratio (SNR) may be insufficient with the use of a celestial source. We can derive a simple expression for the obtainable measurement accuracy as function of the SNR in the following way (Scott and Ryle, 1977). If we measure the pattern in n^2points, each measurement over the angular range $n\lambda/d$ represents an area $1/n^2$of the aperture area and delivers a signal equal to $1/n^2$ of that of the full surface pointed at the source. If this area element has a position error ϵ, it produces a relative component to the signal of $4\pi\epsilon/\lambda n^2$. We combine however measurements from n^2directions to determine the value of ϵ after the Fourier Transformation. For a certain SNR with the source on-axis and equal integration time as each beam point, we obtain for the measurement error

$$\Delta\epsilon = \mathrm{n}\,\lambda\,/\,4\,\pi\,\mathrm{SNR}. \tag{6.3}$$

We will discuss more practicalities later in this chapter. First we want to look at the special problems of holography with the transmitter in the nearfield of the antenna.

6.1.3. Details of the mathematics of nearfield holography

Because we want to derive the complex aperture distribution from the measured nearfield pattern, the inverse Fourier Transformation of Eq. (6.2) will be our point of departure. With reference to Fig. 3.2 and using the direction cosines, introduced in Ch.. 3.7, the expression for the distance r from a point in the aperture to the location of the transmitter at point P becomes

$$r = \{(Ru - \xi)^2 + (Rv - \eta)^2 + R^2(1 - u^2 - v^2)\}^{0.5}, \tag{6.4}$$

which we expand into the series

$$r \approx R - (u\xi + v\eta) + \frac{\xi^2+\eta^2}{2R} - \frac{(\xi^2+\eta^2)^2}{8R^3} - \frac{(u\xi+v\eta)^2}{2R} + \frac{(\xi^2+\eta^2)(u\xi+v\eta)}{2R^2} \tag{6.5}$$

R is the distance from the antenna aperture center to the holography signal source.

Normally for the Fresnel region analysis, the series expansion of r is stopped after the quadratic term, which preserves the first three terms in Eq. (6.5). Here, we shall maintain the next terms as well in order to make an estimate of the error in the approximation. Substitution of Eq. 6.5) into Eq. (6.2) yields

$$F(\xi,\eta) = \frac{e^{-ikR}}{R}\exp\{-ik\,\delta_p(\xi,\eta)\}\int f(u,v)\exp\{ik(u\,\xi+v\eta)\}\,e^{-ik\epsilon}\,du\,dv \tag{6.6}$$

The terms in Eq. (6.5), which are independent of the integration variables, have been brought outside the integral under the variable δ_p. The other terms in higher powers of (u, v) are collected under the variable ϵ. They "modify" the Fourier Transformation of Eq. (6.2).

The first pathlength term

$$\delta_p(\xi,\eta) = \frac{\xi^2+\eta^2}{2R} - \frac{(\xi^2+\eta^2)^2}{8R^3} \tag{6.7}$$

causes a rapidly varying phase variation over the aperture, which can be compensated to a large degree by an axial displacement of the feed (see also Ch. 4.3, Eq. (4.18)). From Ch. 2.2.2, Eq. (2.2.29) we take the relation for the path length variation due to a focus adjustment δf away from the reflector

$$\delta_c(\xi,\eta) = \left\{\xi^2+\eta^2+\left(f-\frac{\xi^2+\eta^2}{4f}+\delta\mathrm{f}\right)^2\right\}^{0.5} - \left\{f+\frac{\xi^2+\eta^2}{4f}+\delta\mathrm{f}\right\}. \tag{6.8}$$

To illustrate the behaviour of these terms as function of the radial aperture coordinate we show here a few figures, calculated with *Mathematica* [Mat.6.1]. Let us first look at the two pathlength terms δ_p and δ_c. The phase errors caused by these are shown in Fig. 6.1 for the practical case of the ALMA antenna measurement, where f = 4.8 m, λ = 3 mm, R = 315 m and the radial coordinate in the aperture runs from 0 to 6 m The "cosine component" of the phase function of Eq. (6.7) is shown in Fig. 6.1 in red, while the phase function of Eq. (6.8) for a choice of δf = 96 mm is superposed in blue. The difference between both terms is shown in the lower plot. The phase error increases its spatial frequency for increasing radial aperture coordinate. The difference plot shows that the residuals are significant. By varying the value of δf in the computation, one can obtain an impression of its influence on structure and magnitude of the difference function. We want to minimise the sum of the two terms (Eqs.(6.7), (6.8)) by choosing an appropriate value of δf. Because of the (ξ,η)-dependence (as shown above), there will be a residual pathlength error. Fig. 6.2 shows this residual pathlength error for two choices of δf [Mat.6.2]. For δf = 96 mm we obtain roughly equal positive and negative errors. If δf = 80 mm is chosen, the error is minimised over the inner half radius, but increases to a larger value towards the edge of the aperture. This remaining pathlength error is a correction to the aperture phase distribution, obtained after the Fourier Transformation of the measured beam pattern. It constitutes a simple algebraic correction to the derived aperture phase.

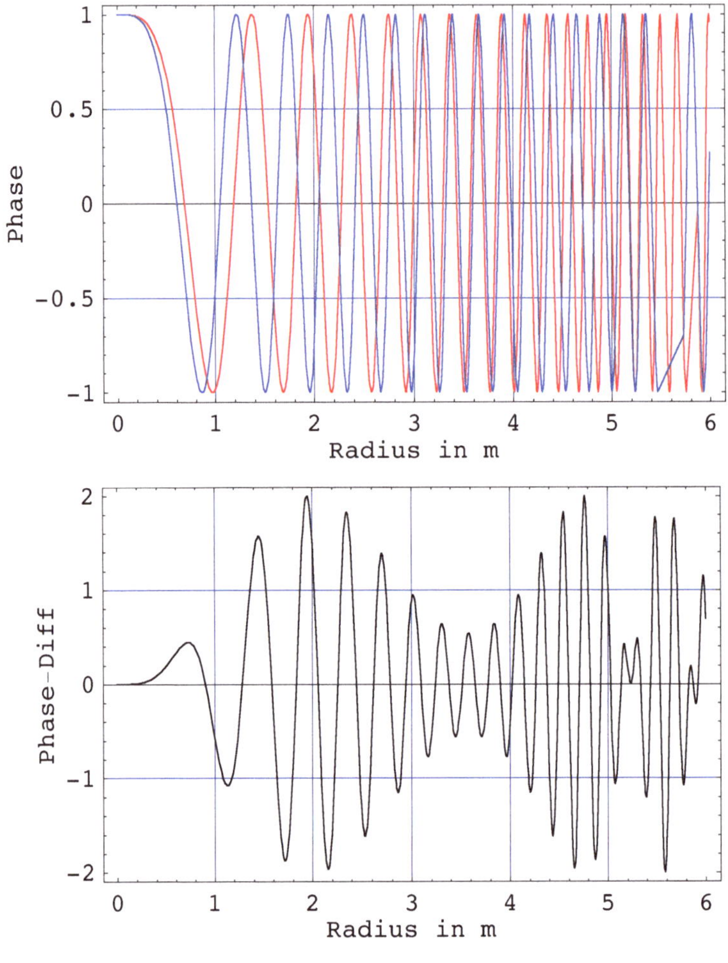

Fig. 6.1. Residual aperture phase for finite distance and axial defocus of 96 mm. Top: Eq. (6.5), red, and Eq. (6.6), blue. Lower curve shows the difference.

The aperture plane must be defined at a convenient location, often halfway between the vertex and the edge of the reflector, where we choose the center of this aperture plane as the origin of the coordinate system. In most antennas there is a significant distance between this plane and the axes of rotation for the movement of the reflector (Fig. 6.3). From this figure we see that there is a "parallax" effect between the used direction cosines (u, v) and those given by the antenna scanning coordinates (u', v'), given by the relations

$$u = u' \left(1 + \frac{\Delta}{R}\right), \; v = v' \left(1 + \frac{\Delta}{R}\right), \qquad (6.9)$$

where Δ is the distance between rotation axis and aperture plane.

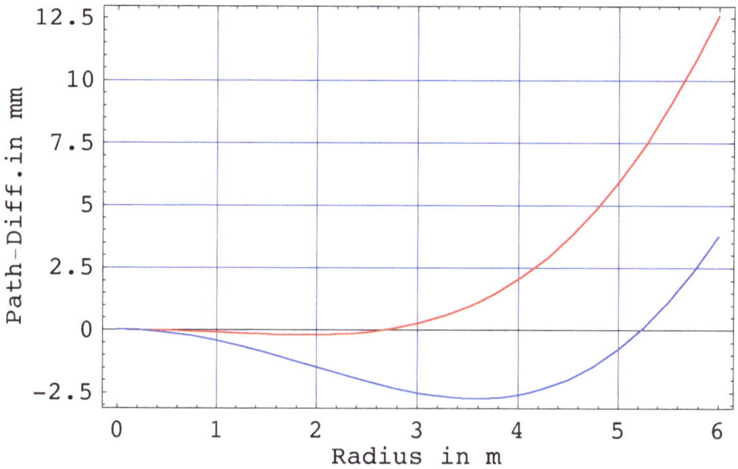

Fig. 6.2. The sum of the terms δ_p and δ_c for values δf = 80 mm (red) and 96 mm (blue). The aperture phase distribution must be corrected according to these residuals.

When we use the scanning coordinates (u', v'), read from the antenna encoders, to calculate the position of the points in the aperture plane (the values for (u, v) in the Fourier integral (Eq. (6.2)), we will overestimate the scale of the aperture map by the factor $(1 + \Delta/R)$. The result of this is that the nearfield correction for each pixel in the map is not evaluated at the correct radius. This causes a pathlength error proportional to the derivative of the nearfield pathlength correction with respect to the radial aperture coordinate. This is illustrated in Fig. 6.4 for the case of the geometry of the ALMA antenna, where $\Delta \approx 3.1$ m, i.e. about 1 percent of the distance R to the transmitter. The pathlength error is significant, causing a surface error as function of radius as shown in the lower part of the figure; its rms value is 18 μm in this case, significant with respect to the required setting accuracy of 10 μm.

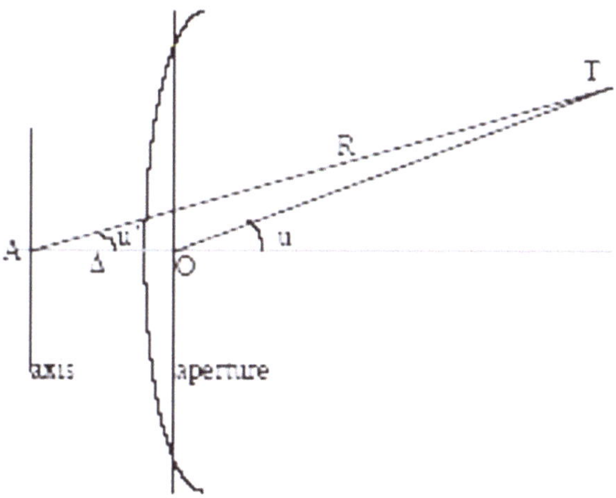

Fig. 6.3. Illustration of the geometry of selected aperture plane and antenna rotation axis.

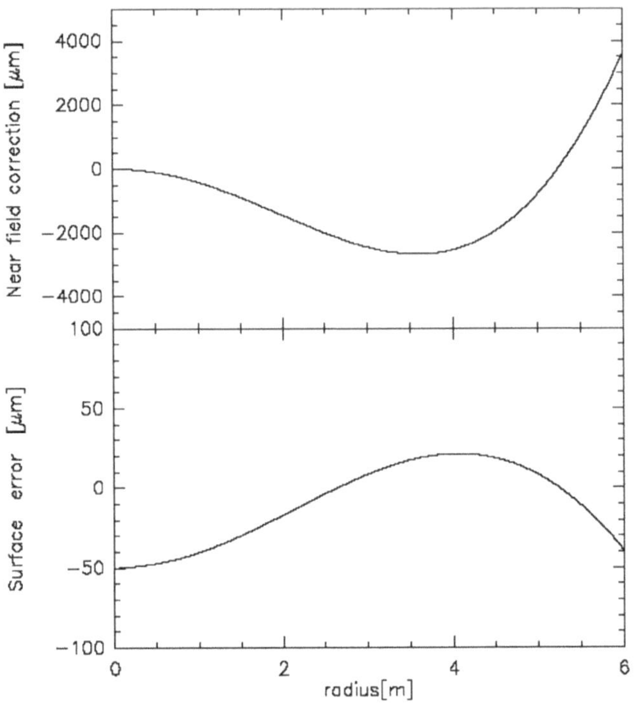

Fig. 6.4. The pathlength error and resulting surface deviation resulting from the difference between the location of the antenna aperture plane and the rotation axes. The rms value of the residual surface error is 18 μm in this example (after Lucas).

It is possible that during the measurement the receiver feed is not located in the optimum refocused position. With the aid of Fig. 6.5 it can be seen that the pathlength error caused by an axial defocus of δz follows from Eq. (6.8) as

$$\delta p_z = \delta z \left\{ 1 - \frac{1 - \frac{\xi^2+\eta^2}{4f^2} + \frac{\delta f}{f}}{\sqrt{\frac{\xi^2+\eta^2}{4f^2} + \left(1 - \frac{\xi^2+\eta^2}{4f^2} + \frac{\delta f}{f}\right)^2}} \right\}, \tag{6.10}$$

while a transverse (lateral) offset by an amount δx will cause a pathlength variation of

$$\delta p_x = \delta x \, \frac{\xi}{f} \left\{ \frac{1}{1 + \frac{\delta f}{f}} - \frac{1}{\sqrt{\frac{\xi^2+\eta^2}{f^2} + \left(1 - \frac{\xi^2+\eta^2}{4f^2} + \frac{\delta f}{f}\right)^2}} \right\}. \tag{6.11}$$

In the reduction process of the holography data, these terms are found by a fit of the measured beam map. The final map of surface deviations is then referred to a position of the feed in the fitted "best-focus" location.

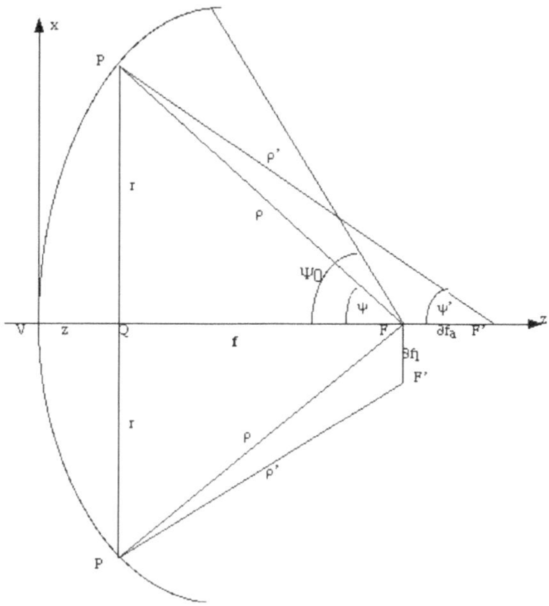

Fig. 6.5. The geometry of axial (upper half) and lateral (lower half) displacement of the feed from the focus of a parabola.

Finally, we look at the higher order terms in Eq. (6.5) containing the integration variables (u,v). They represent a small pathlength error

$$\epsilon = \frac{(\xi^2+\eta^2)(u\,\xi+v\,\eta)}{2\,R^2} - \frac{(u\,\xi+v\,\eta)^2}{2\,R}\,, \tag{6.12}$$

which adds a phase term to the integral of Eq. (6.6) of the following form

$$\exp(-\,i\,k\,\epsilon) \;\approx 1 - \,i\,k\,\epsilon = \tag{6.13}$$

$$1 - i\,k\left\{u\,\frac{\xi(\xi^2+\eta^2)}{2\,R^2} + v\,\frac{\eta(\xi^2+\eta^2)}{2\,R^2} - u^2\,\frac{\xi^2}{2\,R} - v^2\,\frac{\eta^2}{2\,R} - u\,v\,\frac{\xi\,\eta}{R}\right\}.$$

Thus this correction involves the calculation of five additional integrals. Often the additional computational load will not be necessary in view of the required measurement accuracy. If the highest possible accuracy is needed, they should be included.

As an example we take again the ALMA antenna situation mentioned above. When all the integrals of Eq. (6.13) are evaluated, it is found that the contribution of ϵ amounts to 2 μm pathlength over most of the aperture, reaching a value of 5 μm at the edge. This is illustrated in Fig. 6.6. Thus, for a measurement, where the aim is to achieve a measuring accuracy of better than 10 μm, it is advisable to include these terms, if the distance R is relatively small.

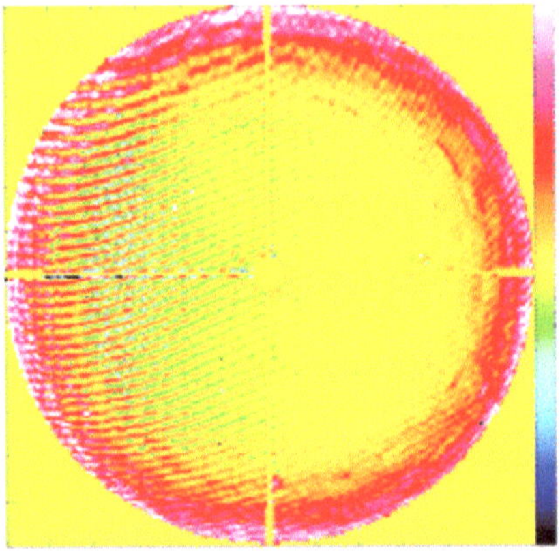

Fig. 6.6. Illustration of the higher order (non-Fresnel, Eq. (6.13)) phase error over the aperture. Over most of the aperture the error is about 2 μm; it raises to more than 5 μm towards the edge.

6.1.4. Aspects of the practical realisation - examples of results

The actual realisation of a holographic reflector measurement will depend strongly on the local situation like the size and frequency range of the antenna, the topographical situation in the neighbourhood of the antenna, the availability of a reference antenna and the accuracy requirement of the measurement. The easiest situation is given by the presence of several antennas as will be the case for an interferometric array. Using some of the strong radio point sources on the sky it will normally be possible to obtain a map of the radiation pattern of one antenna by scanning its beam across the celestial source, while tracking this source with one or more of the other interferometer elements to provide a strong reference signal. In this case the signal source is definitely in the farfield, so no corrections will be needed. Moreover, no special equipment is required as long as the sensitivity of the interferometer is sufficient for the chosen radio source.

Normally however we will be faced with just one antenna to be measured. In that case we must provide for a reference signal from an additional, small antenna. In most cases the sensitivity of the system will not be sufficient for the use of cosmic sources, the only exception being the availability of receiving equipment which is tuned to the strong spectral lines of astronomical maser sources of water-vapour (22 GHz), hydroxyl (OH, 1667 MHZ) or SiO (43 and 86 GHz). Quite a number of antennas have been measured with the aid of beacon signals from communication satellites at frequencies near 12 GHz and 37 GHz. These synchronous satellites are in the farfield of the antenna and their signal is strong and monochromatic, which renders the electronics easier and the data analysis simple.

Where such a solution is not feasible, in particular for the measurement of highly accurate antennas for millimeter and submillimeter wavelengths between 0.3 and 3 mm, there remains only the use of an earthbound transmitter at high frequency, placed at a finite distance from the antenna under test. This transmitter will be well

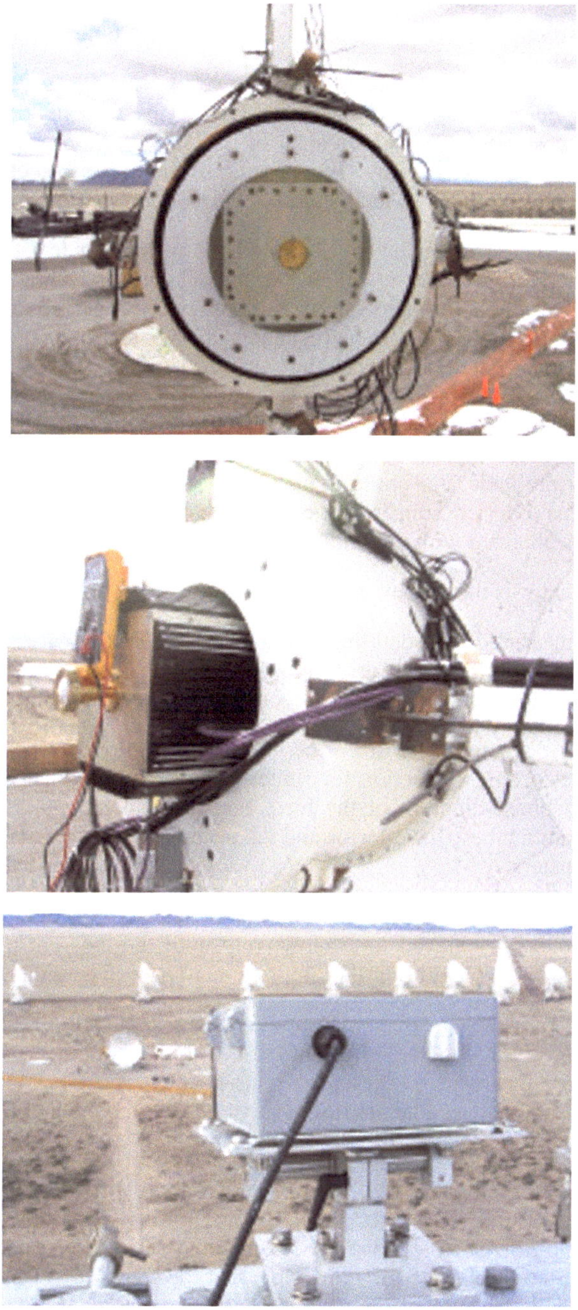

Fig. 6.7. The hardware configuration: signal feed side of the receiver (top), reference feed side of the receiver (middle) and transmitter on top of tower aiming at the antenna in foreground. The other antennas are part of the NRAO Very Large Array in New Mexico.

within the Fresnel region of the aperture. Although it will be easy to obtain sufficient signal to noise ratio, the recorded data will have to be subjected to all the corrective measures described earlier.

A detailed description of the hardware layout, the data collection and the data analysis of the holographic measurement of the ALMA prototype antennas has been presented by Baars et al. (2006). We summarise the major aspects here and present some results as an example of the method.

The antenna has a diameter of 12 m and the reflector panels must be set to an accuracy of 10 μm rms so as to achieve an overall surface accuracy of better than 25 μm. A picture of the holography equipment is shown in Fig. 6.7. The signal source is a monochromatic transmitter at a frequency of 78.9 GHz, located on a 50 m high tower at a distance of 315 m from the antenna. This is at only 0.4 percent of the farfield distance (Eq. (6.1)). The receiver is a dual correlation receiver, located in the apex region behind the primary focus of the reflector. The reference signal is received by a horn - lens combination pointing along the boresight towards the transmitter. Amplitude and phase maps of the antenna beam are obtained by raster scanning. After Fourier Transformation a map of the aperture amplitude and phase distribution is obtained with a spatial resolution over the aperture of about 0.15 m.

In the data analysis the two main operations are:

1. Calibrate measured map data in amplitude and phase, based on regular bore-sight measurements assuming linear drift in amplitude and phase with time. Interpolate the data to a regular grid in antenna based coordinates for easy Fourier transformation.

2. Compute the aperture map from the beam map by Fourier transformation. Apply the geometrical phase correction for finite distance and feed defocus, as well as for the measured phase diagram of the feed. From the aperture phase map, derive the deviations from the best-fit reflector and the necessary adjustments at the positions of the panel adjusters.

3. Apply the corrections to the adjusters and repeat the procedure until the required surface accuracy is achieved, or alternatively until the inherent measurement accuracy has been reached and no further improvement in surface accuracy is obtained.

The result of a holographic measurement and setting of the ALMA prototype antenna is shown in Fig. 6.8. It shows from top to bottom an improving surface accuracy. An adjustment of the panels was done between the maps, based on the results of the one above. The starting error was 53 μm, which could be improved to 14 μm in two full and one partial surface adjustment.

The internal accuracy is normally estimated from the difference between successive measurements. This is illustrated in Fig. 6.9. The rms shape error of the difference map is about 6 μm. There appear to be some relatively large scale differences in the outer areas at the level of 15 - 20 μm. Note that the signal to noise ratio is lower near the aperture edge because of the illumination taper (about -9 dB in this case). If we weigh the data with the taper function, the resulting rms error is just over 4μm.

The holographic method is now being used routinely in the setting of reflector antennas, whereby an accuracy of about one in a million is being achieved.

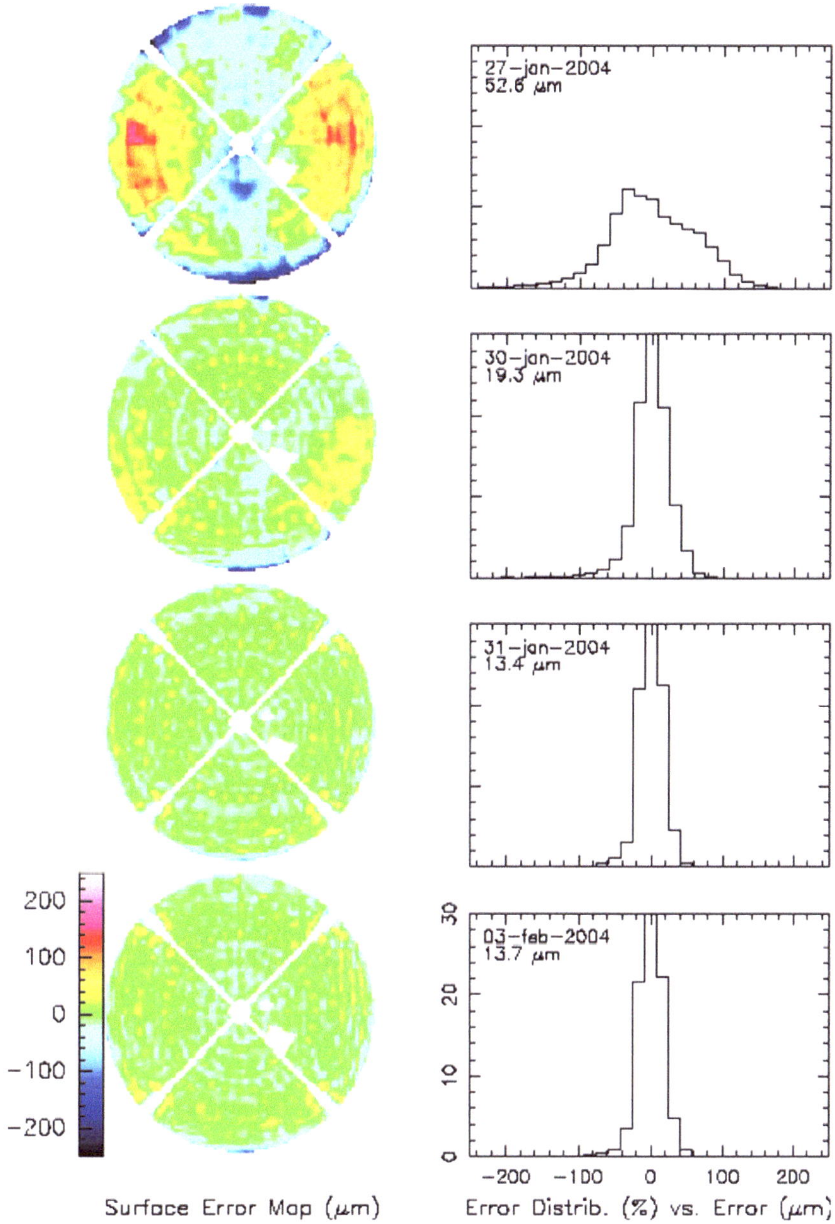

Fig. 6.8. Surface error maps and error distribution of an ALMA prototype antenna. Two full settings were applied between the top three maps and a partial setting before the last map. It appears that the final setting was not significant in view of the measurement accuracy. The white cross and small white area represent the quadripod and a faulty panel.

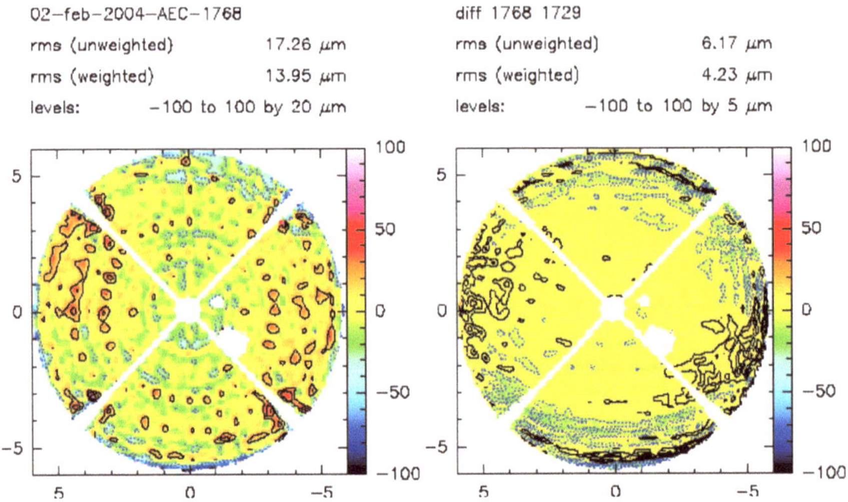

Fig. 6.9. A surface map (left) and the difference between this map and the one made immediately afterwards (right). Here the rms of the surface is 17 μm, while the difference map is about 6 μm. Weighted for the aperture illumination the difference is 4.2 μm.

6.1.5. Alternative methods of reflector shape measurement

Phase retrieval holography

A special version of the holographic method, which only measures the amplitude of the signal, has been developed and successfully used. Here one records two maps at two different settings of the feed along the axis and iteratively derives the phase function over the aperture. Normally one chooses the position of the feed to be about one wavelength at either side of the nominal focus. This way the signal strength will be comparable in both maps. In the manipulation of the data one assumes a starting phase distribution over the aperture, either of a random nature or based on *a priori* knowledge. Normally this would be a smooth function representing the phase pattern of the feed, superposed with a random component. By changing the axial position of the feed a known additional phase function is introduced (see Ch. 3). By an iterative algorithm, originally devised for the analysis of electron microscopy pictures by Misell (1973), the phase function is adjusted to provide a best fit between the calculated and measured amplitude maps in the two focal positions. The method requires a very high signal to noise ratio, but has the advantage of not needing any special hardware.

The required signal to noise ratio is the square of that needed for a direct measurement of the aperture phase function. A detailed analysis has been presented by Morris (1987). Morris et al. (1988) describe a test of phase retrieval holography in the Fresnel region on the IRAM millimeter telescope. The results compared well with the surface map obtained from the full-phase holography. Nicolic et al. (2002) present such measurements of the JCMT submillimeter telescope using astronomical sources. The method has been used widely with satisfactory results. There is however some evidence that the full-phase system delivers data of superior quality.

Shearing interferometer

An original alternative method of surface shape measurement has been presented by Serabyn et al. (1991), in which a Twyman-Green interferometer is used (Born and Wolf, 1970, Ch. 7.5.5). A sketch of the method is shown in Fig. 6.10. The beam from the reflector, after passing through the secondary focus, is led to a tertiary paraboloid P_1, which re-images the primary reflector onto two flat mirrors M_1 and M_2 via a dielectric beam splitter BS. The reflected beams are recombined and brought to the detector in the focal plane via the off-axis paraboloid P_2. Mirror M_2 can be rotated about two axes perpendicular to the incoming wavefront. Because the primary reflector is imaged onto M_2, a rotation of M_2 is equivalent with a change in the pointing direction of the primary reflector. Thus, seen from the detector in the focal point, the fixed mirror M_1 directs a beam towards the source, while the moving mirror M_2 scans a second beam to off-axis positions. In other words, the image mirrors M_1 and M_2 represent the two elements of an interferometer as used in the holographic system described in Ch. 6.1.4. While in standard holography the antenna beam is scanned over the source with the receiver kept on-axis, here the antenna is kept on axis and the focal plane field distribution is sampled off-axis by virtue of the moving mirror M_2. As described in Ch. 3.8, the farfield beam pattern and the focal plane field distribution have the same form (both being the Fourier transformation of the aperture field distribution). Thus this interferometer measurements will deliver the same result as the usual holography interferometry. Because the relative phases

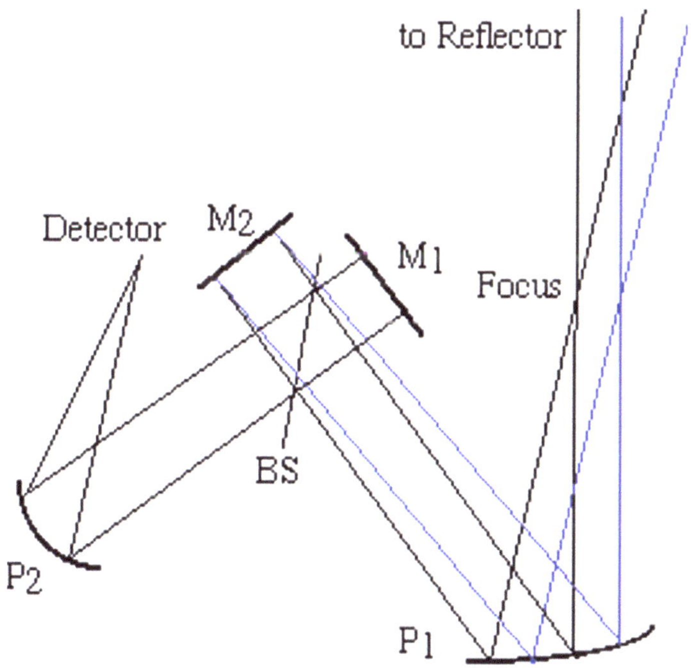

Fig. 6.10. Layout of the shearing interferometer. P_1 and P_2 are off-axis paraboloids; BS the beam splitter and M_1 and M_2 plane mirrors. M_2 can be rotated about two axes perpendicular to the incoming beam; the blue lines indicate an off-axis beam from M_2.

of the two beams are present in the interferogram at the detector, only a single detector is needed and even a broadband device, like a bolometer, can be used. These are frequently used in the millimeter wavelength range, where the planets provide suitably strong test sources. Because the primary reflector is imaged onto mirror M_2 and the wavefronts from a point on the primary surface are sheared laterally by a rotation of M_2, the authors coined the name "shearing holography" for this measurement method. A full description of the method, along with experimental results, is presented by Serabyn et al. (1991).

Optical methods: Photogrammetry and Laser Tracker

For completeness we mention shortly two recent developments of methods which do not use radio techniques, but work with visible light. Both have been used for the setting of accurate reflectors for millimeter wavelengths. While not achieving the accuracy of radio holography, they have some specific advantages which deserve to be mentioned here.

In *photogrammetry* pictures are taken of the reflector surface, which has been outfitted with suitable targets, from a number of directions. From the set of photos a three-dimensional representation of the target locations is formed and hence the reflector shape is recovered. The first application of this method to radio telescopes was a measurement of the 300-ft and 85-ft antennas at the NRAO in Green Bank, WV, in 1962 (Findlay, 1964). The camera was hovered on a helicopter over the antenna. An accuracy of about one millimeter was achieved, i.e. a relative accuracy worse than $1{:}\ 10^5$. The technique has since then been developed to a fully digital

Fig. 6.11. The ALMA prototype antennas are covered with 1080 photogrammetry targets, placed in 17 rings of 24 - 96 targets (white dots).

system and it has been used in the setting of one of the ALMA prototype antennas of 12 m diameter. The camera was placed on a man-lift close to the reflector. This enabled the measurement of the surface in several elevation angle positions. This is of interest to study gravitationally induced deformations in the structure. The accuracy reached is a few times per million and the reflector could be set to approximately 30 μm rms accuracy. Figure 6.11 shows a picture of the reflector with the more than 1000 targets attached.

The original method to measure and set reflector panels was the geodetic "theodolite and tape" method. The use and limitations of this techniques have been described by Greve (1986). With great care a measurement accuracy of somewhat better than 0.1 mm can be obtained. Improvements to the equipment over the years have led to the current use of laser-based theodolites where angle and distance measurement have been combined into one instrument. The so-called "total station" *laser tracker* from Leica is able to track a moving corner-cube mirror target on its path through space. This instrument has been used for the setting of another ALMA prototype antenna. The target was placed at the position of the panel adjusters and readings in real-time enabled the panel to be adjusted, after which the target was moved to the next point. With this instrument it was possible to set the 12-m diameter reflector to an accuracy of about 30 μm.

While these two methods lack the accuracy needed for the setting of the most accurate submillimeter telescopes, where an accuracy of better than 10 μm is required, they offer an attractive alternative for the measurement and setting of antennas of less demanding accuracy.

■ 6.2. Far sidelobes, Gain calibration

6.2.1. Far sidelobes and stray radiation correction

In the first section of this chapter we have seen how the measurement of the antenna pattern in amplitude and phase can be used to derive the shape of the reflector. The method of choice to obtain the amplitude function, rather than the power pattern, is to use an interferometer consisting of the antenna under test and a second antenna which provides a reference signal of constant amplitude and phase by keeping it directed at the signal source. It is obvious that this method can be used for the direct determination of the beam pattern over an arbitrary beam solid area provided the signal to noise ratio is sufficient. A knowledge of as large a portion of the full antenna beam is desired in certain circumstances and the interferometric method often enables us to measure this with the aid of the strongest cosmic radio sources.

In radio astronomy a most serious source of possible error occurs in observations of the atomic hydrogen line at 21 cm wavelength, because the hydrogen in our Galaxy is seen in all directions. The bulk of the radiation is concentrated along the Galactic equator in a band of several degrees width. However, hydrogen is also present at high galactic latitudes, be it at relatively low intensity. Especially the so-called high-velocity clouds (HVC) are objects of great astronomical interest. While observing these, the much stronger radiation from the galactic plane may enter through the far-out sidelobes of the antenna pattern. This so-called *stray radiation* can cause a significant distortion of the HVC's observed line profile and lead to erroneous conclusions regarding the distribution and velocity of the hydrogen in the HVC. The first to draw attention to this was van Woerden (1962), who noticed significant differences between hydrogen spectra from the same direction but observed at different dates. He correctly interpreted this as the result of a variation in the radiation entering through the far sidelobes caused by the orientation of the sidelobe pattern with respect to the brightest regions of the Galaxy and the change in the correction for the velocity of the earth with respect to the source. With the aim of correcting the hydrogen sky surveys undertaken with the Dwingeloo radio telescope, a measurement of the full pattern of the antenna at 21 cm wavelength was undertaken, whereby a level of -60 dB below the main beam needed to be reached (Hartsuijker et al., 1972). In this case the interferometer was formed by the 25-m diameter Dwingeloo telescope and a 7.5 m diameter reference antenna at 100 m distance (Fig. 1.2). Signal sources were the strong cosmic radio sources Cas A, Cyg A and Tau A.

The output of the interferometer is proportional to the product of the voltages from the two elements. By scanning the antenna under test through the direction to the source, while keeping the reference antenna directed at the source, the output signal varies proportional to the *field strength pattern* of the test antenna. The usual power pattern is the square of this and thus a -60 dB power sidelobe level will cause an output voltage of one thousandth of the maximum. The sensitivity of the system allowed to reliably measure sidelobe levels of -55 to -60 dB with respect to the main beam maximum and the entire pattern (within the restrictions of source visibility) was measured; a total of 19000 points over 60 percent of the full sphere. The resulting map is shown in Fig. 6.12. Details can be found in Hartsuijker et al. (1972). In the forward part of the pattern we discern three rings of enhanced radiation centered at θ=30° and ϕ= 60, 180 and 300 degrees, respectively, with a radius of 30° and a width of 3-6°. The level in these rings is about 6 dB above the general sidelobe level in the surrounding areas and about 4 percent of the incoming power is removed from the main beam and scattered into these rings by the tripod supporting the prime focus instrument box. This behaviour is quantitatively in agreement with a theoretical analysis by Rusch et al. (1982). A further discussion of the influence of the feed support on the characteristics of an antenna is presented by Kildal et al. (1988). The rear part of the pattern clearly shows the enhanced level of the spillover ring which lies just outside the geometrical shadow of the reflector rim as seen from the feed at θ = 125°. Normally, the spillover ring is falling on the earth surrounding the antenna. It causes an increase in the system noise temperature, which will be somewhat dependent on the elevation angle of the antenna.

The scatter rings from the feed support legs contribute strongly to the "stray radiation" component in the observations of the 21-cm hydrogen line. Knowledge of the detailed antenna pattern enables one to correct for this if the hydrogen distribution over the entire sky is known. Clearly, for this procedure to work a number of iterations will be necessary, depending on the *a priori* knowledge of the sky brightness distribution. This could be obtained by observing the sky with a "clean" beam as provided by an unblocked aperture like a horn antenna or offset reflector. Hartmann et al. (1996) have carried through the correction of an all-sky hydrogen survey with the Dwingeloo telescope by modeling the antenna pattern on the basis of the measurements presented by Hartsuijker et al. (1972) and the low angular resolution hydrogen line survey by Stark et al. (1992) with the Bell Laboratories 20-ft horn-reflector

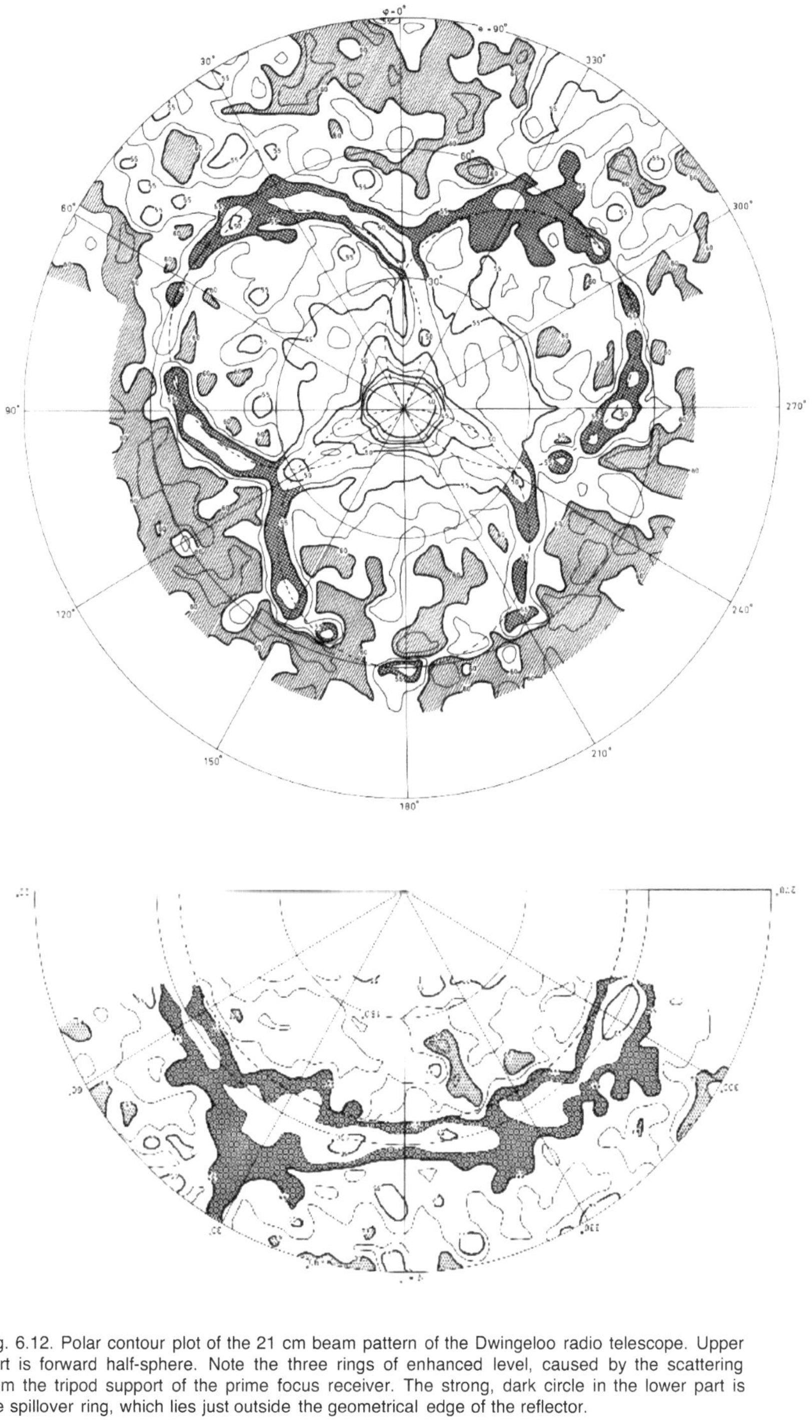

Fig. 6.12. Polar contour plot of the 21 cm beam pattern of the Dwingeloo radio telescope. Upper part is forward half-sphere. Note the three rings of enhanced level, caused by the scattering from the tripod support of the prime focus receiver. The strong, dark circle in the lower part is the spillover ring, which lies just outside the geometrical edge of the reflector.

antenna, which shows a very low and featureless sidelobe structure. Their model of the far sidelobe pattern of the telescope is shown in Fig. 6.13. (When these observations were made, the tripod of the telescope had been replaced by a quadripod.)

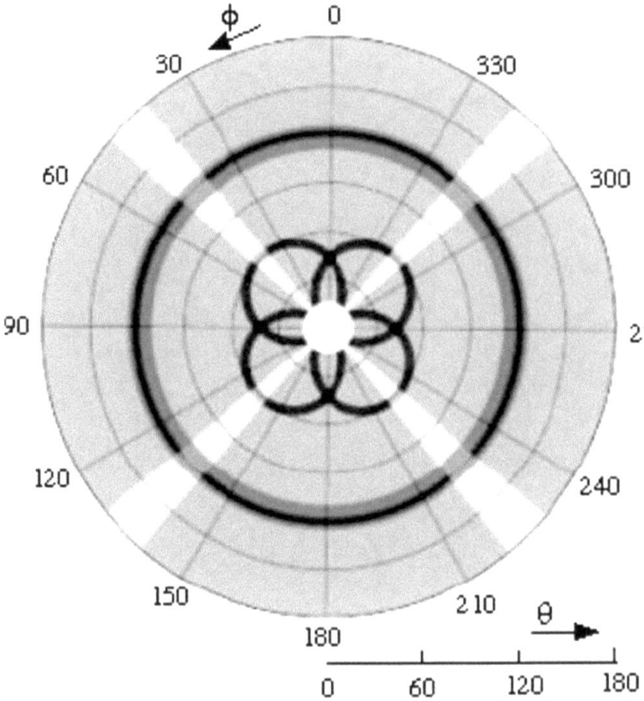

Fig. 6.13. Polar contour plot of a model of the entire beam pattern, derived from the measurements of Fig. 6.12, adjusted to a situation with four support legs, whose shadow forms the white cross and whose scattering cones form the black clover-leave. The dark spillover ring is also visible. The level of the scatter- and spillover-rings is about 6 dB above the surrounding general sidelobe level (after Hartmann).

6.2.2. Absolute gain calibration with an interferometer

We have earlier discussed the need for absolutely calibrated antennas in order to establish a set of calibrator sources on the sky which will then be used to obtain calibrated observations of other objects without the need for a telescope with known gain. Radio astronomers have made good use of the small number of relatively large horn antennas for this purpose, but often the signal to noise ratio is insufficient for accurate results. A powerful method is to observe a source with an interferometer consisting of a relatively small standard gain horn and a large reflector antenna. The gain of the horn can be accurately calculated (Schelkunoff, 1943). The easiest and most accurate way is offered by the existence of an interferometer array with two or more element antennas. The horn antenna is attached to the side of one of the antennas and the receiver is switchable between the horn and the antenna feed. Observing a strong radio source we then measure the output of the interferometer in two situations; one with the horn and the other with the antenna. The second antenna delivers a constant and strong reference signal. After allowing for any difference in loss between the horn and feed output ports (which will have to be measured separately) the ratio of the interferometer output signals is equal to the voltage ratio of the horn and the antenna gains. The gain of the main antenna is then equal to the sum of the output ratio, expressed in dB, and the known gain of the horn. The gain of the horn will typically be some 40 dB less than that of the antenna. While in a total power measurement the horn gain would be too low for an accurate measurement, in the interferometer the output signal will be one percent of that of the two antenna interferometer and still amenable to accurate measurement. With a realistic accuracy of one percent for the gain of the horn, the gain of the antenna can thus be measured to 1-2 percent without the need to know the precise flux density of the source.

This procedure can be repeated for the other interferometer element and in principle be extended to the remaining antennas of a multi element array. Alternatively, the measured antenna temperature due to the source can be measured accurately with a careful thermal calibration method of the receiver system and hence the source flux density can be determined from the known gain of the antenna. This absolutely known flux density can now be used to determine the gain of other antennas without the need for the horn comparison measurement.

If only one antenna is available, the gain of which needs to be determined, one can form an interferometer between it and the standard gain horn and measure the correlated output. The source is also observed with the antenna under test alone. As before, we need to make an accurate calibration of the sensitivity of the receiver systems with the aid of matched loads at different and known temperature to express the output voltage in equivalent antenna temperature at the input of the receiver. Now we have the following relations:

total power measurement: $T_{Ar} \propto S\, A_r,$

interferometer measurement $T_{Ai} \propto S\, (A_r\, A_h)^{0.5},$

where S is the (unknown) flux density of the source, A_r and A_h the absorption areas of the reflector (to be determined) and the horn (known), respectively. From this follows

$$A_r = 4\left(\frac{T_{Ar}}{T_{Ai}}\right)^2 A_h. \qquad (6.14)$$

The proportionality sign hides some terms related to the electronic gains of the two different systems, which can however be accurately determined in the laboratory.

These methods have been applied by Welch and colleagues to calibrate the antennas of the BIMA array (Welch et al., 1996) and to establish some absolute planetary brightness temperatures at 11 and 3 mm wavelength to an accuracy of 1-2 percent (Gibson et al., 2005; Gibson and Welch, pers. com., Sep. 2005). This is a great step forward in the calibration quality in the millimeter wavelength domain. The goal of a five percent flux calibration for the ALMA submillimeter array may be reached by applying these methods.

■ 6.3. Chromatism, Off-set aperture

6.3.1. Chromatism - "baseline ripple"

As became clear in the discussion of the stray radiation, it is possible that radiation reaches the feed through more than one path. When this happens the pathlength differences upon reaching the detector will cause interference, the intensity of which will be frequency dependent. In spectral line observation, where the total bandwidth is analysed into several hundreds of separate channels, this interference can lead to a periodic ripple, normally called *baseline ripple*. Weinreb (1967) was the first to draw attention to this fact. While baseline ripple can also be caused by mismatches within the receiver electronics, we limit ourselves here to shortly describe the causes of baseline ripple which are connected to the antenna and its feed. A thorough treatment of this case has been presented by Morris (1978). Most important is the backscattering of radiation by the feed and its surrounding mounting plate. Some of this radiation is reflected again by the reflector and via the feed support structure to re-enter the feed, where it will give rise to the interference mentioned above. The amplitude of the ripple is proportional to the wavelength. Possible paths are illustrated in Fig. 6.14. Most serious is path A, the specular reflection from the center region of the reflector.

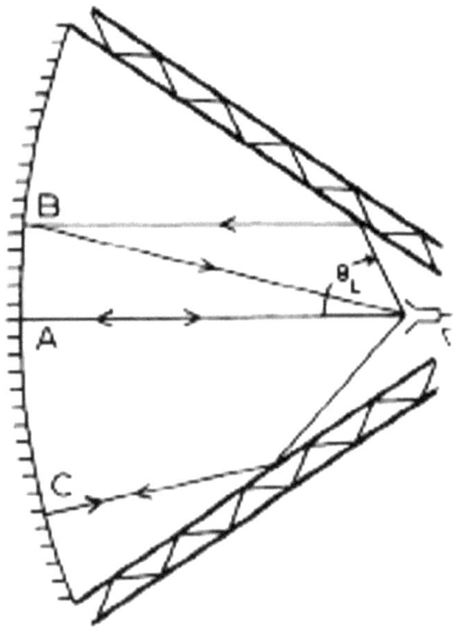

Fig. 6.14 - Sketch of three paths along which radiation scattered from the feed and its surroundings can be returned to the feed and cause interference with the directly received signal. (After Morris, 1978)

Depending on their structural layout, the support legs can contribute significantly to the ripple. The "periodicity" of the ripple caused by path A will be c/2 f, where c is the velocity of light and f the focal length of the antenna. Thus this will be 10 - 30 MHz for typical antennas of 10 - 50 m diameter. In the case of a Cassegrain configuration, we need to replace f by m·f with m the magnification factor. Thus with typical values of $10 < m < 20$ the ripple will be in the lower MHz range. The ripple due to path A can be significantly suppressed by combining two observations taken with the feed axially defocussed at plus and minus $\lambda/8$ from the focal point. The $\lambda/2$ total path difference between the doubly reflected rays will effectively cancel the ripple effect. This is illustrated in Fig. 6.15. The multi-path reflections via the support legs will not be completely cancelled this way. Methods to reduce the chromatism are described by Padman and Hills (1991).

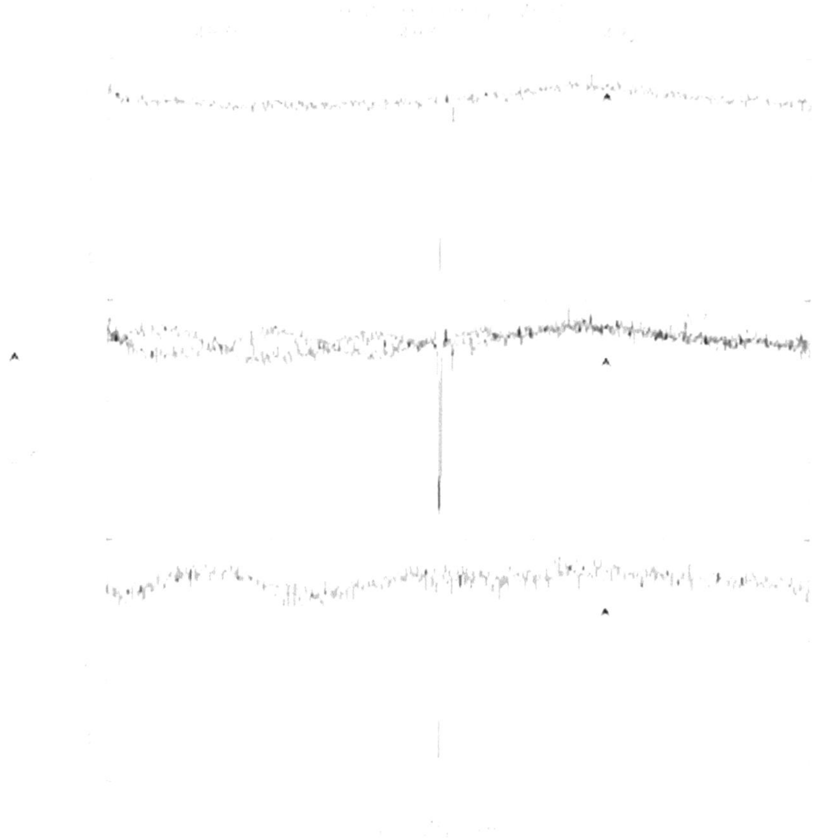

Fig. 6.15. Spectrum of an absorption line in the presence of baseline ripple (lower frame). In the middle frame there are two observations whereby the feed has been axially defocused by + and - $\lambda/8$. Adding these to form the upper trace shows that the ripple with 300 km/s "period" has been suppressed. The wide ripple in the top frame is due to another multipath interference. (Pillai, MPIfR)

In the design of modern radio telescopes the avoidance or minimisation of these multiple reflections has been included in the design effort. As Morris showed, the following measures can greatly contribute to the suppression of baseline ripples:

i) path A: avoid reflections on structures near the feed by covering these with absorbing material. Of course the reflection on the feed itself cannot be avoided. Minimise the backscatter from the central region of the reflector to the feed by a "splash plate", directing the radiation outwards. In a Cassegrain system this can be achieved effectively by shaping the central area of the subreflector into a cone. The size and angle of the cone are chosen such that the radiation is directed to the main reflector just outside the central hole which provides access to the secondary focus, where the feed is located.

ii) paths B and C; here it is important to avoid specular reflection on the support legs by shaping the legs with a sharply pointed wedge towards the main reflector. This will reflect the radiation away from the direction of the main reflector.

iii) in a Cassegrain system there will be some edge diffraction on the subreflector. This diffracted radiation will enter the feed directly and for a perfectly circular secondary reflector the feed will be on the caustic of the diffracted rays. Thus relatively large interference effects with the reflected rays can be expected. This can be significantly reduced by making the subreflector slightly un-round and/or serrated. The un-roundness need only be of the order of a wavelength; thus a "rough" edge with serrations of a few millimeter would also be effective.

All three precautions were taken in the design and construction of the 30-m millimeter telescope of IRAM. This telescope exhibits low baseline ripples as illustrated in Fig. 6.16. A small, fast residual ripple of 42 m equivalent length is still visible. It most likely originates in the electronics, but could be caused by a reflection path in the telescope of 42 m length.

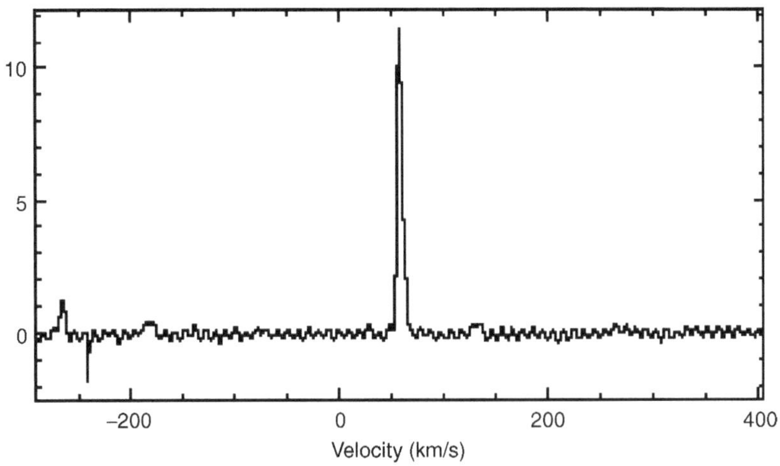

Fig. 6.16. Example of a spectral line observation with the IRAM 30-m telescope. This is a "raw" spectrum, showing a very flat baseline with a small remaining fast ripple, caused by a 42 m long "double reflection" path in the telescope structure. (Mauersberger, IRAM)

6.3.2. Unblocked aperture - Offset antenna

We have illustrated above the importance of the far sidelobes for the correct interpretation of observations of extended radiation. An example is the measurement of the Cosmic Microwave Background (CMB) radiation, where the stray radiation might easily be as large as the true signal in the main beam. An antenna without features in the sidelobe pattern like feed support scatter rings and spillover ring would be welcome. This can be achieved to a significant degree by arranging the mirrors and feed in such a way that the main reflector aperture is not blocked by any structural elements. These so-called offset antennas have received great attention in communication technology, both for satellite on-board antennas and the ubiquitous small satellite dishes for direct TV reception. A trivial example of an unblocked aperture is

Fig. 6.17. The Bell-Laboratories 7-m Offset-antenna at Holmdel, NJ, USA. Subreflector is to the left; The person is entering the feed- and receiver-cabin under the main dish. (Photo by the author, Courtesy AT&T)

the horn antenna, which is well understood theoretically. Indeed the beam of such an antenna is so "clean" that Penzias and Wilson (1965), using the 20-ft folded horn antenna at the Bell Laboratories could confidently determine an excess of 3 K in antenna temperature when looking at blank sky. For this discovery of the Cosmic Microwave Background radiation they were awarded the Nobel Price in 1978.

The "open Cassegrain" antenna was described originally in a paper by Cook et al. (1965), also of the Bell Laboratories. This was followed by the actual construction of an offset antenna of 7 m diameter with a highly accurate surface of about 0.1 mm rms error (Chu et al., 1978). The antenna (Fig. 6.17) was used both for radio astronomy at 115 GHz and for propagation experiments with Comstar beacons at 19 and 29 GHz. Extensive measurements of the beam characteristics were made which confirmed the theoretical predictions.

In the following years the construction of clear aperture radio telescopes was considered occasionally during design phases, but did not result in actual realisation. Firstly, the unsymmetrical structure leads to the need for additional material and hence cost and, secondly, there was the fear that such structures would behave less homologously and would not achieve the required performance over the full elevation range. The collapse of the 300-ft NRAO transit telescope in Green Bank in 1988 changed the situation decisively. When NRAO received funds to replace the telescope by a fully steerable antenna of some 100 m diameter, it was decided to build a clear aperture telescope. The overriding argument was based on astronomical

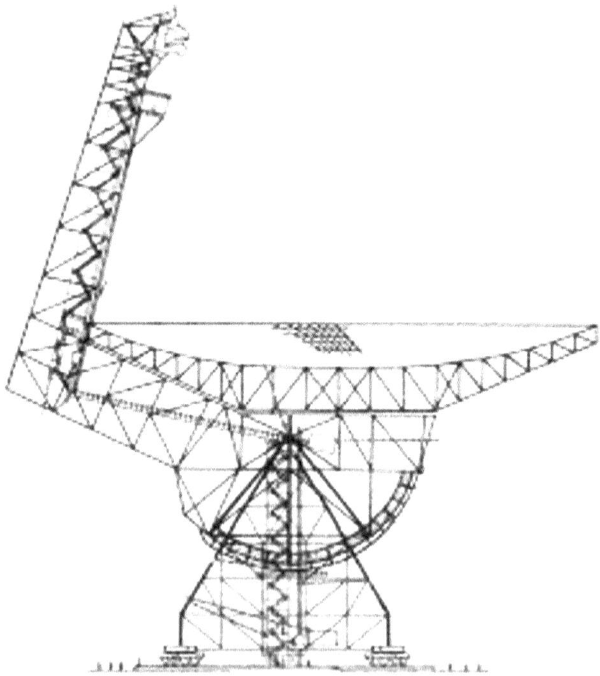

Fig. 6.18. Cross section through the Green Bank telescope (GBT), an offset antenna of 100 m diameter. (NRAO/AUI)

requirements, as illustrated in the earlier sections. The technical challenges were formidable and indeed the design and construction have met with considerable difficulties. These have been resolved and the antenna is operational. A photograph has been shown in Fig. 1.7 and Fig. 6.18 presents a sketch of the cross section through the structure.

The electromagnetic analysis of an offset antenna is far from trivial. The loss of symmetry causes several complications and since the original paper by Cook the subject has received much attention. While these studies fall outside the scope of this book, it is useful to mention a few of the major aspects of the electromagnetic behaviour of the offset antenna. The illumination function, emanating from a circular horn aperture will not be uniform over the reflector aperture. For this reason the reflector is made slightly elliptical with the larger dimension in the plane of the feed support (see Fig. 1.7). Numerous studies of the beam parameters, polarisation behaviour and noise characteristics have appeared and form a rather complete body of data which can be used to evaluate the expected performance of the offset antenna. For a review see Rudge and Adatia (1978).

A major, and generally decisive, argument in the choice of an offset antenna is the absence of structured sidelobes, which is a direct result of the unobstructed aperture. Thus, the antenna will have no blocking and hence exhibit a somewhat higher gain than a symmetrical antenna. Typical blocking losses of a symmetrical system are 5 - 20 percent. As illustrated above in Section 6.2.1, the very low and regular sidelobe structure is of great importance for the suppression of wide-angle interfering radiation. The offset antenna exhibits a larger instrumental polarisation due to the oblique incidence of the feed radiation pattern onto the reflector. Considerable efforts have been devoted to special feed designs to minimise this effect. While this is outside the scope of our discussion, we refer to the collection of papers on Reflector Antennas, edited by Love (1978), where these aspects are discussed.

■ 6.4. Atmospheric fluctuations and dual-beam observing

6.4.1. Introduction

Depending on the frequency of operation, radio astronomy observations, as well as communication channels, suffer under the influence of the earth's atmosphere. While short-wave radio in the frequency range from about 5 - 30 MHz employs the reflective characteristics of the ionosphere to realise trans-horizon communication, transmissions in the GHz range are hampered by the absorption and scattering of the constituents of the troposphere, mainly oxygen and water vapour. In this section we pay some attention to the latter and we describe a method whereby radio astronomy observations can be made less sensitive to the effects of the atmosphere.

The interaction between the troposphere and an electro-magnetic wave passing through it is two-fold: refraction and scattering on the one hand and attenuation on the other. The latter is connected to the self-radiation of the atmosphere through Kirchhoff's law. The atmospheric parameter describing the interaction is the complex permittivity ϵ, which can be written as $\epsilon = \epsilon_1 - i\,\epsilon_2$. Here the real part ϵ_1 determines the refractive index n of the medium, while the imaginary part ϵ_2 is related to the absorption α. A general relation between the refractive and absorptive parts of the permittivity was established by Kramers (1927) and Kronig (1928), now known as the Kramers - Kronig relations and written as

$$\epsilon_1(\nu) - \epsilon_\infty = \frac{2}{\pi} \int_0^\infty \frac{x\,\epsilon_2(x)}{x^2 - \nu^2}\, dx\,, \tag{6.15}$$

$$\epsilon_2(\nu) = \frac{-2\,\nu}{\pi} \int_0^\infty \frac{\epsilon_1(x) - \epsilon_\infty}{x^2 - \nu^2}\, dx\,. \tag{6.16}$$

The integration extends over the entire frequency domain; ϵ_∞ is the permittivity for the frequency tending to infinity. A proof of these relations can be found in e.g. Landau-Lifshitz (1960), Vol.8, §62. We see that we could calculate the refractive index at one frequency, if we knew the absorption over the entire frequency domain, and vice versa. The following relations also hold:

$$n = \sqrt{\epsilon_1} \qquad \text{and} \qquad \alpha = 2\pi\,\epsilon_2/\lambda, \tag{6.17}$$

where we assume that the relative permeability $\mu_r = 1$.Because in the atmosphere the refractive index n is only slightly larger than one, it is convenient to introduce the *refractivity* N, defined as

$$N = (n-1)\,10^6 \approx \tfrac{1}{2}\,(\epsilon_1 - 1)\,10^6\,. \tag{6.18}$$

In the microwave and millimeter wavelength region the tropospheric refractivity is determined mainly by oxygen and water vapour. These molecules exhibit spectral lines in this frequency region due to rotational transitions. We will not discuss these further here apart from mentioning that oxygen has a broad absorption band between 50 and 60 GHz as well as a sharp line at 118 GHz. Water vapour lines occur near 22 GHz , 183 GHz and higher frequencies in the submillimeter region. The atmospheric absorption is shown in Fig. 6.19 for a relative water vapour density of 1 percent. There is also a non-resonant component to the absorption and the pressure broadened spectral lines cause significant absorption outside the transition frequencies. Normally one will avoid the resonant frequencies for observation or transmission.

It is of interest to obtain an expression for the refractivity which shows the dependence on the temperature, pressure and water vapour content of the atmosphere. Such a semi-empirical expression has been found by Smith and Weintraub (1953):

$$N = \frac{77.6}{T}\left(P + \frac{4810\,e}{T}\right), \tag{6.19}$$

where T is the absolute temperature (K), P the total atmospheric pressure (mBar) and e the water vapour pressure (mBar). The numerical coefficients are derived from experimental data. These are not dimensionless; thus the formula is dimensionally correct.

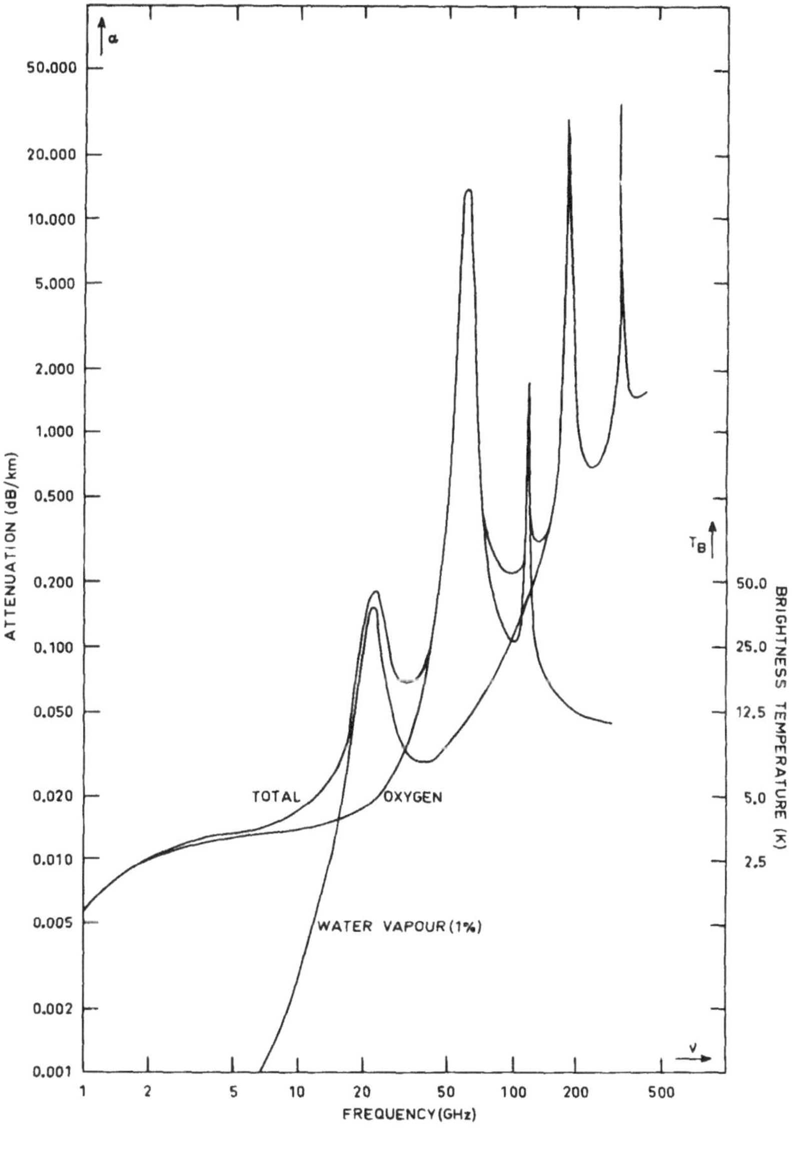

Fig. 6.19. Attenuation in the horizontal surface layer of the troposphere for oxygen, water vapour (assumed relative vapour density of 1%) and the total atmosphere. Right hand scale is atmospheric brightness temperature looking straight up for a standard atmospheric model.

In the following sections we shall not present an exhaustive discussion of the propagation in the troposphere, but limit ourselves to some of the main results in view of application to observations and consequences for observing strategies. A more complete treatment, in particular in relation to interferometric observations, can be found in the work by Thompson, Moran and Swenson (2001), where many relevant references are given.

6.4.2. Atmospheric emission and attenuation

As stated earlier, the main absorptive constituents of the atmosphere are oxygen and water vapour, each of which have absorption bands in the millimeter and submillimeter wavelength region. Considerable effort has been devoted to constructing a propagation model of the atmosphere and to derive the resulting absorption as function of frequency. Major contributions are due to Liebe (1989) and also Pardo et al. (2001) for the submillimeter wavelength region. While the oxygen contribution is rather constant and similar over the world, the contribution of water vapour is highly dependent on the local weather situation. Generally the models assume an exponential decrease of the oxygen and water vapour concentration with height. The scale height is defined as the height where the concentration has decreased to a value 1/e = 0.37 of the surface value. For oxygen the scale height is typically 8 km, while that of water vapour is about 2 km. Thus it is clear that a significant decrease in the influence of water vapour can be achieved by locating the telescope on a high and dry site. This has indeed been done with most dedicated millimeter telescopes.

The zenith attenuation (often called opacity from the usage in optical astronomy) as shown in Fig. 6.19 is connected to the atmospheric self-radiation through Kirchhoff's Law. If we denote the zenith opacity by τ_0 we can write for the antenna temperature at the antenna terminals

$$T_A(A) = T_A(0)\, e^{-\tau_0} + T_{\mathrm{atm}}\,(1 - e^{-\tau_0}), \tag{6.20}$$

where $T_A(0)$ is the antenna temperature at the top of the atmosphere and T_{atm} the effective temperature of the atmosphere. The zenith opacity τ_0 will vary according to the changes in the water vapour density and must normally be measured at regular intervals. This is often done by performing a "tipping scan" measurement, whereby the opacity is determined from the antenna temperature as measured over a large zenith angle range (without background sources). The air mass m_{atm} (the total column mass of the atmosphere), defined to be unity towards the zenith, increases to a first approximation as the secans of the zenith angle, whereby the atmosphere is considered as a set of plane parallel sheets. For large zenith angles a more accurate formula might be needed, whereby the curvature of the atmospheric layers is taken into account. Rohlfs and Wilson (1996) give a power series, valid to 4 air masses (zenith angle of 75 degrees) with an error of less than 1 percent, of the form

$$m_{\mathrm{atm}}(z) = -0.0045 + 1.00672 \sec z - 0.002234 \sec^2 z - 0.0006247 \sec^3 z\,. \tag{6.21}$$

Performing a tipping scan, we easily see from Eq. (6.20) that the measured antenna temperature as function of the zenith angle z will be

$$T_A(z) = T_{\mathrm{atm}}\,(1 - e^{-\tau_0 \sec z}). \tag{6.22}$$

Plotting the logarithm of the measured signal against sec z (the air mass along the line of sight) delivers the zenith opacity from the slope of the curve. Care should be taken that the measurement is not corrupted by varying radiation received in the sidelobes of the antenna, which will pick up significant amounts of ground radiation for large zenith angles. For this reason the scan is normally stopped at about 70 degrees zenith angle. The opacity can also be found from the change in measured antenna temperature from a sufficiently strong radio source as function of zenith angle. In this case we must however assume that the atmospheric opacity is constant over the long duration of the measurement, because the zenith distance varies by no more than 15 degrees in one hour. Therefore, the tipping scan has been adopted at many observatories as a routine method to quickly determine the atmospheric opacity.

6.4.3. Atmospheric refraction

Apart from the attenuation suffered by the radiation on its path through the atmosphere, the finite refractivity will cause the rays to be refracted and delayed with respect to transmission through a vacuum. In the discussion of pointing in Chapter 5 we already mentioned the influence of refraction on the apparent pointing direction towards the source. The atmospheric delay is normally not important for single dish observations of cosmic radio sources. It does play a significant role in interferometry, in particular Very Long Baseline Interferometry, where the element antennas are so far apart that the tropospheric structure is completely uncorrelated between the antennas. Delay is also of high importance in timing applications as with the Global Positioning System of satellites. Aspects of this have been treated by Herring (1992) and Rüeger (1996). A detailed treatment of this aspect falls outside the scope of this book.

We will however summarise the important aspects of refraction, as it significantly influences the pointing of the antenna, Refraction is defined as the bending of an electro-magnetic wave on its path through a medium of variable index of refraction. As shown in Eq. (6.19), the refractivity is dependent on the temperature and pressure of the atmospheric constituents. Thus tropospheric refractivity will be dependent on height, decreasing from its surface value to essentially zero at the edge of the earth atmosphere. Under the assumption that the atmosphere can be described by a set of plane parallel layers of slowly decreasing refractivity, an elementary analysis (as for instance nicely presented by Smart in his Textbook on Spherical Astronomy, 1962) leads to the surprisingly simple relation for the angle of refraction

$$\Delta z = (n_0 - 1)\tan z\,, \tag{6.23}$$

where the angle z is the zenith angle and n_0 is the index of refraction at the level of measurement. The apparent zenith distance of the source is Δz smaller than the true zenith angle. Clearly, the assumption of a plane parallel atmospheric structure is expected to be valid only for small zenith angles. If we describe the atmosphere by a number of parallel spherical layers, the resulting angle of refraction contains an additional term proportional to $\tan^3 z$ with a coefficient about three orders of magnitude smaller than $n_0 - 1$. Several workers have made accurate measurements of the refraction at radio wavelengths and have found that the simple formula of Eq. (6.23) can be used with sufficient precision up to zenith angles of 80 degrees. It should be noted that such measurements are not trivial, because one must be able to separate any elevation dependent pointing error of the antenna from the refraction.

Large millimeter wavelength telescopes now operate with beamwidths of the order of 10 arcseconds, while the refraction at 45° elevation is about 1 arcminute. It is thus important to determine the instantaneous surface refractivity with good accuracy in order to apply refraction corrections to the antenna pointing at the level of arcseconds. Fortunately much attention has been given recently to improving the original Smith and Weintraub formula (Eq. 6.19), in particular in view of the GPS system. With "best average" coefficients, as suggested by Rüeger (2002), the formula for the surface refractivity can be written as

$$N = (n_0 - 1)\,10^6 = 77.6890\,\frac{p_d}{T} + 71.2952\,\frac{e}{T} + 375463\,\frac{e}{T^2}\,, \qquad (6.24)$$

where p_d is the partial pressure of the dry atmosphere, e the partial pressure of water vapour, both in hPa, and T the temperature in K. With typical sea level values $p_d = 1000$, $e = 10$ and $T = 288$ we find $\Delta z = 66 \tan z$ arcseconds; on a high and dry mm-telescope site we might have $p_d = 700$, $e = 2$ and $T = 270$ and hence $\Delta z = 44 \tan z$.The difference illustrates the importance of applying real-time refraction corrections determined from local temperature and pressure measurements.

In the following sections of this chapter we shall discuss the fluctuating component of the atmospheric attenuation and self-radiation. Clearly, these relatively small-scale fluctuations in the atmospheric constituents will also influence the refraction angle. For instance a doubling of the water vapour from 2 to 4 hPa in the example above changes the refraction coefficient from 44 to 46 arcseconds. Very large variations of refraction of the order of tens of arcseconds over time spans of up to one minute have been observed. The first to identify these was Altenhoff (Altenhoff et al., 1987) through careful analysis of pointing measurements with stable telescopes and beamwidths of less than one arcminute. These large fluctuations have received the unfortunate name of *anomalous refraction* although there is basically nothing anomalous about it. They are caused by large blobs of humid air passing through the beam of the antenna at relatively small distances from the aperture. A simple argument provides an order of magnitude description of the phenomenon. Consider a wedge of humid air passing with the wind over the aperture in the Fresnel region of the antenna. Assume a gradient in the wet part of the refractivity $\Delta N_w = 5$ over an assumed thickness of the wedge of 100m The pathlength variation will be

$\Delta l = 100\,\Delta N_w\,10^{-6} = 0.5$ mm. This pathlength difference between opposite edges of the antenna of width D will cause a change in angle of arrival of $\Delta\alpha = \Delta l / D$. Over the width of an aperture of 30 m this amounts to an angle of about 3.5 arcsecs.

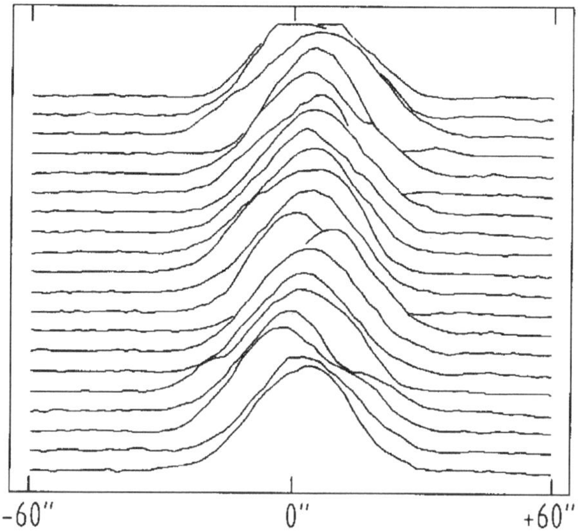

Fig. 6.20. Example of "anomalous refraction". A series of consecutive fast scans of a point source, time stepping from front to rear. Position shifts of 5-10 arcseconds in a time lapse of less than 10 seconds are clearly visible.

This is similar to what is being observed. An example of the phenomenon is shown in Fig. 6.20, taken from Altenhoff et al. (1987). Changes in "pointing" of the order of 20" over a time period of a minute are apparent. The effect has been seen with the newer and accurate millimeter telescopes at a number of different sites and appears to be a general feature of a not fully stable atmosphere.

6.4.4. Signal fluctuations due to atmospheric turbulence

The dynamic behaviour of meteorological processes causes fluctuations in the atmospheric temperature, pressure and water vapour content, both on a temporal and spatial scale. They are caused by wind, solar radiation, frontal activity, etc. The properties of these fluctuations are described by the *statistical theory of turbulence*. We are interested in the influence of the fluctuations on the propagating electro-magnetic wave. This has been treated in great detail by Tatarski in his monograph "Wave propagation in a turbulent medium" (1961). We summarise here some of the major results of this work which are of direct interest to the subject at hand.

The central statistical quantities of any parameter of the fluctuating medium are the mean value and the correlation function. The latter describes to what extent the fluctuations at different position or times are related. A stationary random process has a constant mean value and its correlation function depends only on the time difference. Obviously the atmosphere is not strictly stationary. For such processes

Kolmogorov introduced the concept of stationary *first increments,* where $f(\tau) = f(t+\tau) - f(t)$ is stationary for some interval τ. The correlation function of the increments, called the *structure function*, is written as

$$D_f(\tau) = \overline{\{f(t+\tau) - f(t)\}^2} \,.$$

It characterises the intensity of fluctuations of $f(t)$ over periods smaller than or comparable to τ.

Considering the three-dimensional spatial turbulent structure of the atmosphere, we can apply the concept of structure functions over distances where the turbulent field can be considered *locally homogeneous and isotropic*. We obtain a similar structure function for the fluctuations between points a distance r apart:

$$D_f(r) = \overline{\{f(r+r_1) - f(r_1)\}^2} \,. \tag{6.25}$$

The turbulent flow is characterised by the inner and outer scale of turbulence, l_0 and L_0, respectively. At the inner scale the turbulent energy is converted to heat; in the atmosphere $l_0 \approx 1$ cm. The outer scale depends on the Reynolds number of the flow and falls in the region of tens to hundreds of meters.

For a distance r between two points with $l_0 \ll r \ll L_0$ it has been shown by Kolmogorov (1941) that the structure function for velocity or pressure fluctuations is given by (see also Tatarski, 1961, Ch. 2)

$$D(r) = a\,(\epsilon\, r)^{2/3} \,. \tag{6.26}$$

This is the so-called "two-thirds law", where ϵ is the energy dissipation rate per unit mass and time. The theory can be extended to the behaviour of the refractive index in the turbulent atmosphere. Without going into the details, we state that a structure function of the same form is found for the fluctuations in the refractive index.

$$D_n(r) = C_n{}^2\, r^{2/3} \,, \tag{6.27}$$

where the structure constant for atmospheric refractivity is dependent on the gradient of the refractivity and the outer scale of turbulence L_0. It is given by

$$C_n = a\, L_0{}^{2/3}\, \Delta \mathrm{n}, \tag{6.28}$$

where a is a constant of order unity and $\Delta n = \Delta N \,.\, 10^{-6}$ the refractive index fluctuation. Typical values for ΔN in the lower atmosphere are 2, decreasing to 0.1 - 0.3 N-units at 2 - 3 km altitude. Due to the large range of L_0, determined by the local topography and general weather pattern, the value of C_n varies appreciably. It has been derived from meteorological measurements and radio wave propagation experiments and the range found is $10^{-8} < C_n < 4.10^{-7}\ m^{-1/3}$, thus typically $C_n \approx 10^{-7}\ m^{-1/3}$.

The turbulent structure of the atmospheric refractivity causes fluctuations in the amplitude and the phase of an electro-magnetic wave traveling through the medium. Also, the varying optical depth of the atmosphere, both temporally and spatially, will cause a fluctuation in the self-radiation of the atmosphere, adding a noise component to the measured antenna temperature. The details of the interaction have been treated extensively in Tatarski's monograph (1961, Chs. 7 - 8) and will not be repeated here. It is found that the interaction of a wave of wavelength λ with a turbulent path of length L is mainly determined by turbulent regions of a size $\sqrt{\lambda L}$, under the assumption that $l_0 \ll \sqrt{\lambda L} \ll L_0$, which is normally the case in the troposphere and the microwave region. The turbulons of this size contribute mostly to the amplitude fluctuations and the correlation distance of the amplitude fluctuations, after traversing the turbulent layer is of the order $\sqrt{\lambda L}$. While over a horizontal path we can assume C_n to be constant on average, it will change with height and for a slanted path to a satellite or radio source the effect of C_n must be integrated along the path.

Tatarski obtains the following expressions from his analysis:

i) the mean square logarithmic amplitude fluctuations $\overline{\chi^2}$:

$$\overline{\chi^2} = \ln\left(\frac{A_0 + \Delta A}{A_0}\right)^2 = 0.56\ k^{7/6} \int_0^L C_n{}^2(h)\, h^{5/6}\, dh \tag{6.29}$$

ii) the correlation coefficient $b_A(\rho)$ of the amplitude fluctuations in two points a distance ρ apart:

$$b_a(\rho) = 1 - 2.37\left(\frac{k}{L}\right)^{5/6} \rho^{5/3}\,, \quad \left(l_0 \ll \rho \lesssim \sqrt{\lambda L}\right) \tag{6.30}$$

iii) the structure function $D_S(\rho)$ of phase fluctuations between two points a distance ρ apart:

$$D_S(\rho) = 2.91\, k^2\, \rho^{5/3} \int_0^L C_n(h)^2\, dh, \quad \left(l_0 \ll \sqrt{\lambda L} \lesssim \rho\right). \tag{6.31}$$

Here A_0 is the average amplitude, $k = 2\pi/\lambda$ the wavenumber, L the total path through the atmosphere. For a slanted path with a zenith angle θ_z the pathlength will be $L = h \sec\theta_z$, where h is the effective height of the turbulent atmosphere. The dependence of C_n on the height in the troposphere has been derived by Fried and Cloud (1966) as

$$C_n(h)^2 = 4.2 \cdot 10^{-14}\, h^{-1/3} \exp(h/h_0)\ ,\ h_0 = 3200\, m, \tag{6.32}$$

with h in meters and $C_n{}^2$ in $m^{-2/3}$. The quantity h_0 is called the scale height.

Using these expressions, we can make an estimate of the expected phase and amplitude fluctuations at the terminals of our antenna. From the definition of the structure function (Eq. (6.25)) it is clear that the relative phase fluctuation $\Delta\Phi$ between two points a distance ρ apart is given by the square root of $D_s(\rho)$. Note that the phase fluctuation increases with the 5/6th power of the "baseline" ρ, thus slightly less than linearly, and inversely proportional to the wavelength. If we use Eq. (6.31) with Eq. (6.32) for a wavelength $\lambda = 3$ mm, a distance $\rho = 100$ m, we obtain $\Delta\Phi = 0.6$ radian $\approx$ 35 degrees (Fig. 6.21) [Mat.6.3]. If we move the location of the antenna from sea level to 3000 m altitude, the fluctuations decrease by about a factor 2.

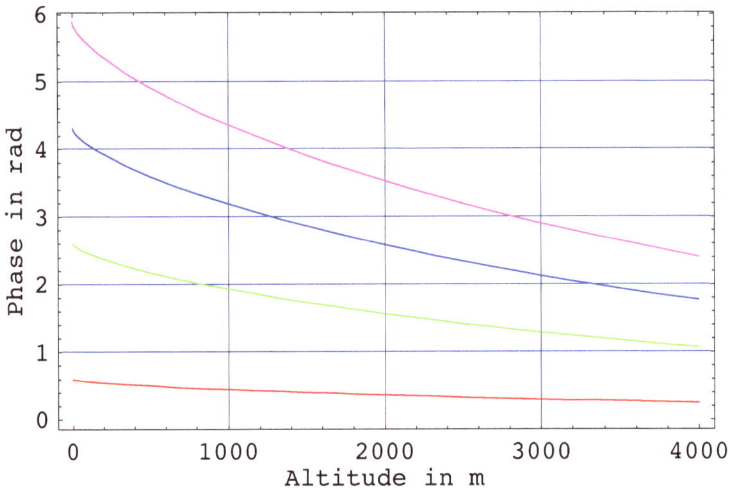

Fig. 6.21. Phase fluctuation as function of the altitude between two points 100-step 500-1600 m apart, from bottom to top.

For the subject of this chapter it is of more interest to look at the amplitude fluctuations, because these are directly related to the fluctuations in the self-emission of the atmosphere by virtue of Kirchhoff's law:

$$\eta = \alpha\, B(\nu, T), \tag{6.33}$$

where η and α are the coefficient of emission and absorption, respectively; B(ν,T) is

Planck's radiation law (or its Rayleigh-Jeans approximation, where appropriate). The antenna measures fluctuations in power, which are proportional to the square of the amplitude fluctuations. Using Eq. (6.29) with Eq. (6.32) and integrating along L to infinity, we obtain for the mean square logarithmic fluctuation in atmospheric brightness temperature ΔT_B:

$$\overline{\ln\left\{\left(\frac{T_{B0}+\Delta T_B}{T_{B0}}\right)\right\}^2} = 4\,\overline{\chi^2} = 1.5\,.\,10^{-8}\,k^{7/6}. \qquad (6.34)$$

The atmospheric average brightness temperature must be calculated for a standard model of the atmosphere. Atmospheric physicists have made an extensive effort to establish such a standard model. For wavelengths in the centimeter and millimeter range the model of Liebe (1989) is widely accepted. Note that the brightness temperature, and hence the fluctuating component of atmospheric emission and absorption is strongly determined by water vapour. This is the main reason why antennas for work in the mm-wavelength region are located at high sites in dry areas of the world. Also, it should be mentioned that the above discussion assumes a clear atmosphere without influence of clouds and frontal activity. The turbulence discussed here occurs in the quiet atmosphere because of small fluctuations in temperature, pressure and humidity. In Fig. 6.22 we show the relative power fluctuations as function of wavelength as computed from the above equations [Mat.6.4]. We see a rather strong increase with decreasing wavelength, which is caused mainly by the increased opacity of the atmospheric water vapour. It should be borne in mind that this curve is valid for the frequency regions outside the pressure broadened absorption lines (see Fig. 6.19), where the above approach breaks down. It is also clear that the relative fluctuation level decreases slowly with increasing altitude, about a factor two for 5000 m. We should however remember that the actual atmospheric brightness temperature also

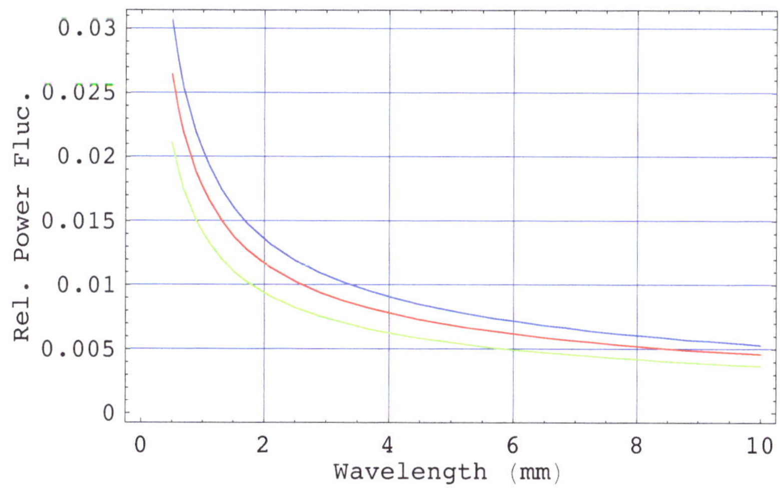

Fig. 6.22. The relative power fluctuation due to atmospheric turbulence as function of wavelength. The curves are for altitudes of 0 (blue), 2000 (red) and 4000 m (green), respectively.

decreases with increasing height. With the assumed scale height of $h_0 = 3200\,m$ it would decrease to 20 percent at 5000 m altitude compared to the sea level value. Thus moving the antenna to 5000 m altitude would diminish the atmospheric fluctuations by an order of magnitude at the short mm wavelengths.

From Fig. 6.19 we can read the atmospheric brightness temperature as function of wavelength. Multiplying these with the relative power fluctuation of Fig. 6.22 yields the expected fluctuation at the antenna terminals due to the clear atmosphere. As an example, at 3 mm wavelength the atmospheric brightness temperature is about 50 K for a relatively dry atmosphere at sea level. We expect a fluctuation level of about 1 percent, or 0.5 K equivalent antenna temperature. This is significant compared to the receiver noise fluctuation, which for a receiver noise temperature of 50 K, a bandwidth of 100 MHz and integration time of 1 second would be 5 mK.

6.4.5. Observing methods to cancel atmospheric fluctuations

The "information" collected by a radio telescope has the characteristic of noise, which is normally not discernible from the inherent thermal noise of the antenna-receiver system. The signal is generally broad-band, continuous radiation. The only exception is the presence of spectral lines of cosmic molecules and atoms, which appear at discrete frequencies, but otherwise are noise-like as well. The signals in communication channels are modulated in a particular fashion and therefore easier to separate from the background noise. In both cases the signal-to-noise-ratio (SNR) must be sufficient to reliably detect the information. While the information in a communication channel continuously changes with time, the radio astronomy signal normally does not vary its character over reasonable time spans. Thus the radio astronomer can increase the SNR of his observation by collecting the signal over a long time.

We omit a detailed treatment of the operation of a radiometer here. Excellent descriptions can be found in, for instance, Kraus (1966) and Tiuri (1964). The important radiometer equation, which determines the SNR can be written as follows

$$\Delta T_{\mathrm{rms}} = m \, \frac{T_{\mathrm{sys}}}{\sqrt{\tau B}} \,, \tag{6.35}$$

where ΔT_{rms} is the rms noise fluctuation at the output terminals, expressed in Kelvin, due to a system with total noise temperature T_{sys}, operating with bandwidth B (Hz) and an integration time τ (sec). The factor m is of order one and accounts for the detailed configuration of the receiver system and the observing mode used to collect the signal. For a simple total power receiver m = 1. It is clear that collecting radiation over a long time decreases the noise fluctuation, hence increases the SNR, proportional to the square root of the total integration time. In this equation, the total system noise temperature is the sum of the instrumental (receiver) noise temperature T_R and the antenna temperature T_A due to the antenna and the source (see Ch. 5.2). If we integrate until $\Delta\, T_{\mathrm{rms}} \ll T_A$, we can reliably determine T_A. This assumes that the receiver system is absolutely stable, i.e. that T_{sys} and the gain of the electronics G_{rec}

are constant over time. Gain stability in an electronic system with active components is difficult to achieve and moreover gain fluctuations cannot be countered by longer integration. Without recourse to special methods, the achievable SNR will normally be determined more by gain fluctuations than system noise. Fortunately gain fluctuations tend to be slow with a frequency spectrum falling off as ν^{-2} or steeper.

It was realised by Dicke (1946) that gain fluctuations could be effectively suppressed if the receiver would be switched between the antenna terminals and a constant reference noise source at a frequency well above that of the dominant gain fluctuations. The difference signal of antenna and reference would thus not suffer from gain fluctuations and the necessary SNR could be achieved by a sufficiently long integration. The cost of this method is a loss in sensitivity of a factor two (m = 2 in Eq. 6.35), because the signal from the sky is only observed half of the time. To achieve the same SNR, the observation will take four times as much time compared to the total power system. Nevertheless, the so-called Dicke-receiver has found wide application in radio astronomy.

The Dicke scheme works best if the reference noise temperature is equal to the antenna temperature, in which case the gain fluctuations cancel perfectly. One way to approach this situation is to use a wide-beam "sky horn", pointing in the general direction of the antenna. A next step is to place two feeds in the focal plane of the antenna, looking with similar beams at adjacent patches of sky, several beamwidths apart. This is a very effective method to observe weak point-like sources. When the telescope is scanned across the direction of the source, the output of the radiometer will show an S-shaped trace of double amplitude as the source first traverses the "signal" beam and then the "negative reference" beam. Actually, this layout was first used in surveys for point sources (Conway et al, 1965, Davis M.M., 1967) to increase the reliability of detections. These observations were carried out at relatively long cm-wavelengths, where the influence of the atmosphere is weak.

It occurred to Conway (1963) that this "dual-beam" method might be effective in the cancellation of the stronger atmospheric fluctuations to be expected at short cm- and mm-wavelengths. The argument is that, while the two beams are well separated in the farfield, they will partially overlap in the nearfield region. At the short wavelengths, this region will extend to well above the height of the tropospheric fluctuations (see Eq. (6.1) and thus the beam overlap will be considerable, resulting in an effective cancellation of tropospheric fluctuations. The author (Baars, 1966, 1970) made a detailed study of the beam overlap and demonstrated the efficacy of the method with observations at several wavelengths. Because this observing method has become an essential aspect of millimeter wavelength radio astronomy, we will summarise the main results now.

Millimeter telescopes are generally of the Cassegrain type. The relatively small size of the secondary reflector, typically between 0.5 and 1 m diameter, makes it possible to attach these to a nutator mechanism, which enables the subreflector to be mechanically moved between different positions. This effectively switches the beam of the antenna between two neighbouring points on the sky, removing the need for a second feed to realise the reference beam. The nutating subreflector had earlier been introduced on infrared telescopes for the same reason.

This method is not restricted to the observation of point sources, which will be seen by only one beam at a time. In an important paper Emerson et al (1979) have demonstrated that a dual-beam mapping of an extended source, which is essentially a differential map of the object, can be restored to deliver a reliable representation of

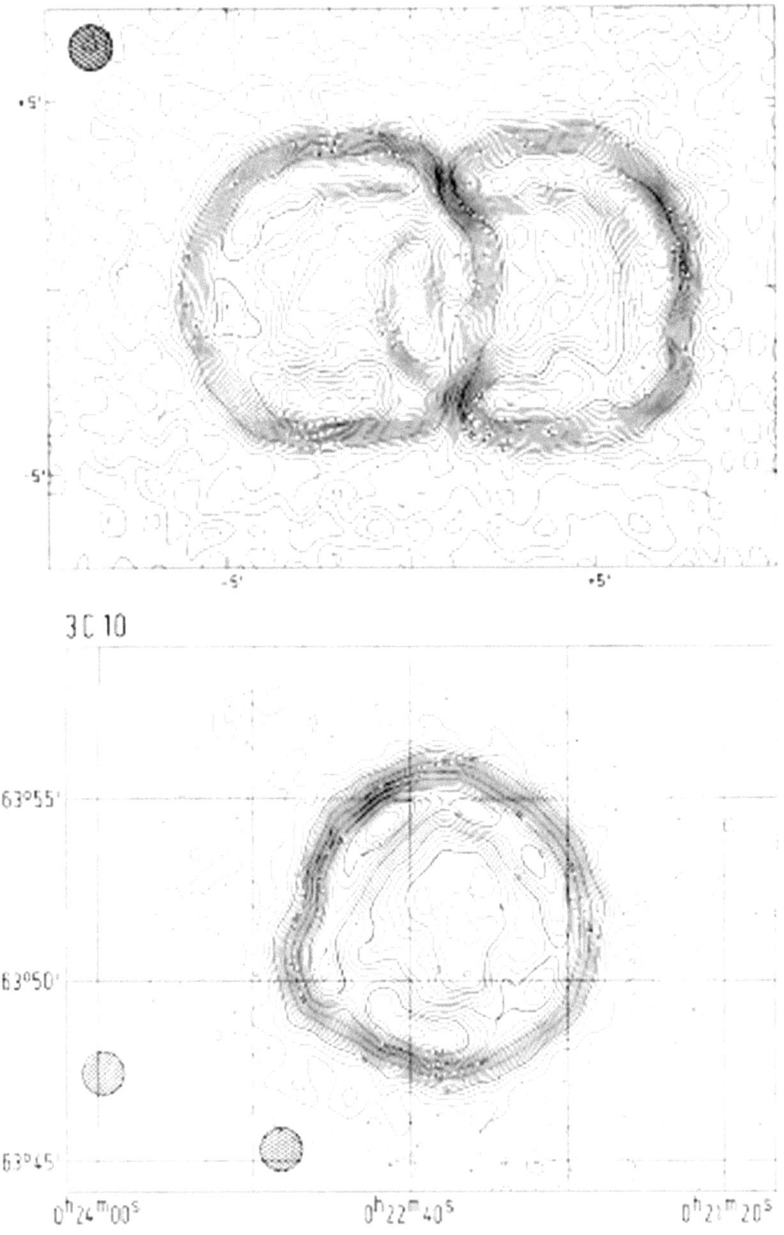

Fig. 6.23. Dual-beam observation of an extended source (upper plot) and the derived map obtained with the EKH-algorithm (lower plot). The beam separation and size is indicated in the lower left hand corner of the lower plot.

the brightness distribution of the source. This "EKH-algorithm" is now widely used in dual-beam observations of extended objects. We show here an example from their paper (Fig. 6.23); the mapping of a supernova remnant 3C10 of 9 arcminutes size with a dual-beam system of 5.5 arcminute beam separation. The upper plot shows the difference output, where the left and right output overlap. The combined result is shown in the lower plot.

6.4.6. Beam overlap in the Fresnel region of the antenna

The necessary mathematical preparations have already been made in Chapter 3 and 4, where we calculated the antenna beam in the Fresnel region and for a laterally offset feed. We need to calculate the beam patterns for a number of distances from the aperture and from those compute the percentage of beam overlap as function of the distance. We shall not carry out these computations here in detail. As illustration of the different shapes of the nearfield patterns, two are shown in Fig. 6.24 for distances from the aperture of 1/128 and 1/160 of the farfield distance, respectively. They involve inserting the nearfield term of Eq. (3.37) into the radiation integral for lateral defocus as given in Eq. (4.33).

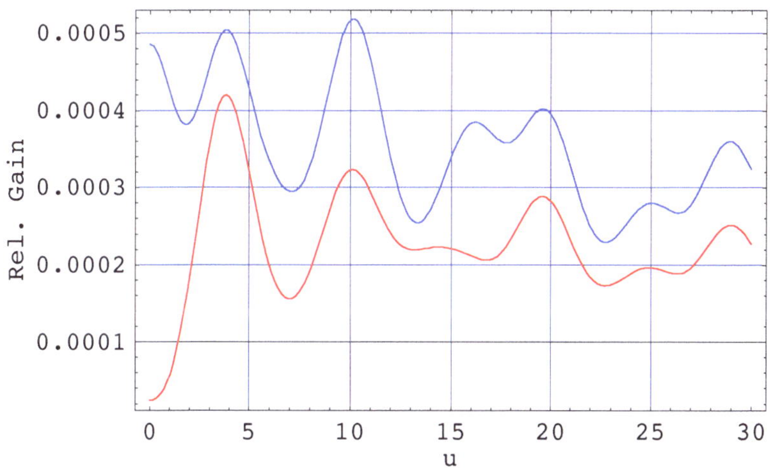

Fig. 6.24. Nearfield radiation function at a distance of 1/128 (blue) and 1/160 (red) times the farfield distance. Note the zero axial value of the red curve.

By calculating the pattern for a series of distances from the aperture we can derive the shape of the beam as function of this distance R. We find that close to the aperture the radiated power flows in a tube with a diameter roughly equal to that of the aperture. At a certain distance the beam widens to the angular extent of the farfield beamwidth. As soon as the beam becomes divergent, the diffraction effects which cause the sidelobe structure of the farfield pattern become visible. For a uniformly illuminated aperture the divergence sets in at a distance of $R_r = d^2 / 2\lambda$, which is called the Rayleigh distance. Rayleigh (1891) noted the significance of this distance in his study of pinhole photography. The Rayleigh distance is one fourth of the farfield distance R_f, defined in Eq. (6.1). With a tapered aperture illumination the transition to the diffraction beam occurs at a somewhat smaller distance. Having established the beam cross-section as function of distance for both slightly displaced beams, we can now calculate the beam overlap, defined as the ratio of the power flowing through the common area of both beams to the power flow through a single beam. An example of such a calculation is shown in Fig. 6.25. After an initial linear decrease the beam overlap becomes weaker with increasing distance. For a 12 m diameter antenna operating at 1 mm wavelength the Rayleigh distance R_f = 72 km. Considering that tropospheric fluctuations occur mainly in the lower 3-4 km, we see that the beam overlap in that region is about 90 percent for a beam separation of 2 HPBW and 75 percent for 4 HPBW separation.

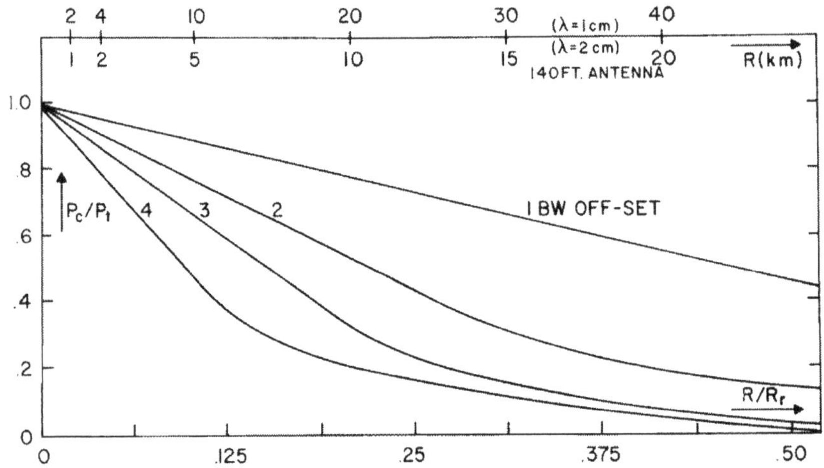

Fig. 6.25. The normalised beam overlap as function of the distance from the aperture in terms of the Rayleigh distance R_r with the offset of the second beam in HPBW as parameter. The top scale indicates real distance for a 43-m diameter antenna and wavelengths of 1 and 2 cm.

It is thus clear that we might expect a significant cancellation of those atmospheric disturbances which are small in size compared to the beam and fluctuate slower than the beam-switching cycle time, which is typically 5-10 Hz. As we saw above, the true structure of the atmospheric fluctuations is described by an ensemble of turbulons with varying scale length and intensity, distributed over the line of sight. The achievable suppression of fluctuations will depend on the scale length and the speed with which the turbulons move through the beam. The method has proved its success and observations at short centimeter and millimeter wavelengths invariably employ a switched dual-beam method. An illustration of the improvement is shown in Fig. 6.26, taken from the work by the author (Baars, 1966). Finally, it should be mentioned that the best results are obtained with a symmetric switching procedure. In that case the sidelobe structure (coma-lobe) of the two beams will be identical leading to a more complete cancellation of the atmospheric fluctuations. Thus placing the two beams symmetrically off-axis is preferable to the on-axis/off-axis configuration.

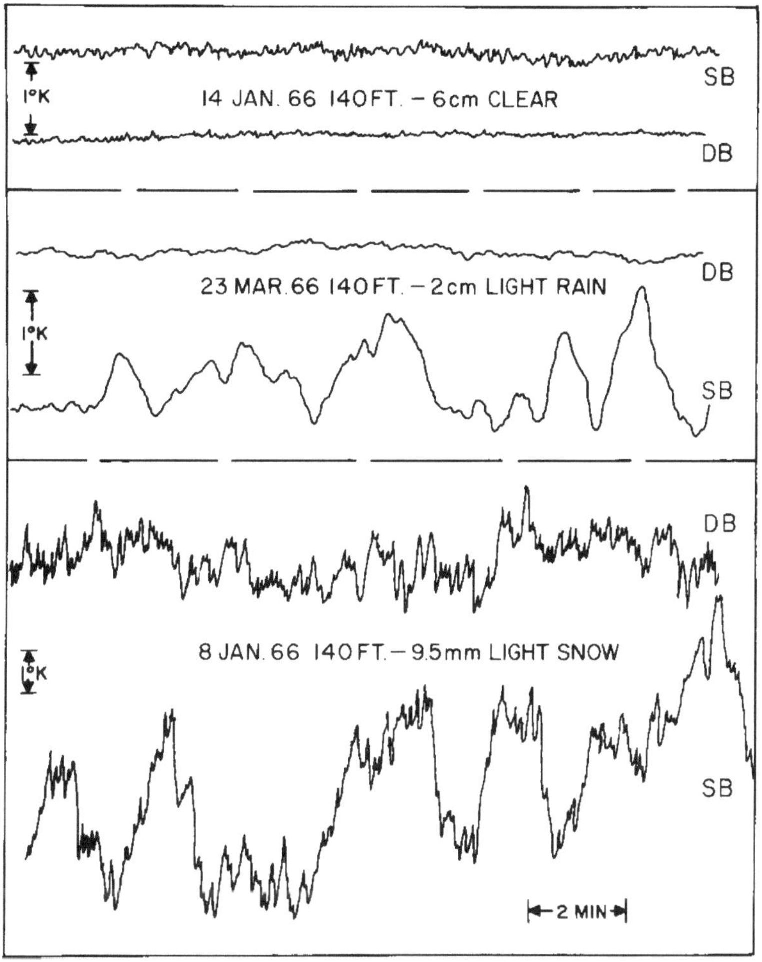

Fig. 6.26. Traces of 30 minutes duration of blank sky observed in single- (SB) and dual-beam (DB) modes at three different wavelengths and under differing weather conditions. The antenna temperature scale is indicated on the left. Note the large improvement in baseline stability of the DB scans. Even in absolutely clear weather at 6 cm wavelength an improvement is clearly present.

We conclude here the discussion of the more practical aspects of the calibration and use of large reflector antennas for radio astronomy and satellite communication. We have barely touched on the performance of existing instruments. In the following chapter we present the main features of the design and performance of a number of important radio telescopes. There we highlight original design aspects and choice of new composite materials. These antennas reach a beamwidth of about 10 arcseconds at their shortest wavelength, indicating a need for 1 arcsecond pointing accuracy. Thus the design must aim at achieving this along with producing a high accuracy reflector surface.

6.5. The *Mathematica* routines

```
Mat .6 .1 - near field path length error;
Clear[δf]; R = 315; f = 4.80; δf = .096;
λ = 0.003; k = 2 π / λ;
dp = c^2 / (2 R) - c^4 / (8 R^3);
dc = (c^2 + (f - c^2/(4 f) + δf)^2)^0.5 - (f + c^2/(4 f) + δf);
php = Cos[k dp]; phc = Cos[k dc];
Plot[{php, phc}, {c, 0, 6}, PlotRange → All, Frame → True,
 GridLines → Automatic, FrameLabel → {"Radius in m", "Phase"},
 PlotStyle -> {{RGBColor[1, 0, 0]}, {RGBColor[0, 0, 1]}}]
Plot[php - phc, {c, 0, 6}, GridLines → Automatic,
 FrameLabel → {"Radius in m", "Phase-Diff"}, Frame → True]
```

```
Mat .6 .2 - path length and defocus;
Clear[δf]; R = 315; f = 4.80; λ = 0.003; k = 2 π / λ;
dp = c^2 / (2 R) - c^4 / (8 R^3);
dc = (c^2 + (f - c^2/(4 f) + δf)^2)^0.5 - (f + c^2/(4 f) + δf);
Plot[Evaluate[
  Table[ 1000 (dp + dc), {δf, 0.080, 0.096, 0.016}], {c, 0, 6},
  PlotRange → All, Frame → True, GridLines → Automatic,
  PlotStyle -> {{RGBColor[1, 0, 0]}, {RGBColor[0, 0, 1]}},
  FrameLabel → {"Radius in m", "Path-Diff.in mm"}]]
```

```
Mat .6 .4 - relative atmospheric amplitude fluctuation;
λ =.; k = 2 π / (λ / 1000); h0 = 3200;
cc = 4.2 10^-14 h^(-1 / 3) Exp[-h / h0];
am2 = 0.56 k^(7 / 6) Integrate[cc * h^(5 / 6) , {h, el, ∞}];
dT = Exp[2 Sqrt[am2]];
Plot[Evaluate[Table[dT - 1, {el, 0, 4000, 2000}],
  {λ, 0.5, 10}, Frame → True, GridLines → Automatic,
  FrameLabel → {"Wavelength (mm)", "Rel. Power Fluc."},
  PlotStyle → {{RGBColor[0, 0, 1]},
    {RGBColor[1, 0, 0]}, {RGBColor[0, 1, 0]}}]]
```

```
Mat .6 .3 - atmospheric phase vs altitude
λ = 0.003; k = 2 π / λ; h0 = 3200; ρ = .;
cc = 4.2 10^-14 h^(-1 / 3) Exp[-h / h0];
ph = Sqrt[2.91 k^2 ρ^(5 / 3) Integrate[cc, {h, ht, ∞}]];
Plot[Evaluate[Table[ph, {ρ, 100, 1600, 500}],
   {ht, 0, 4000}, Frame → True, GridLines → Automatic,
   FrameLabel → {"Altitude in m", "Phase in rad"},
   PlotStyle → {{RGBColor[1, 0, 0]}, {RGBColor[0, 1, 0]},
      {RGBColor[0, 0, 1]}, {RGBColor[1, 0, 1]}}]]
```

References

Altenhoff, W.J., J.W.M. Baars, D. Downes and J.E. Wink, Observations of anomalous refraction at radio wavelengths, *Astron. Astrophys.* **184**, 381-385, 1987.

Baars, J.W.M., Reduction of tropospheric noise fluctuations at centimeter wavelengths, *Nature* **212**, 494, 1966.

Baars, J.W.M., *Dual-beam parabolic antennae in radio astronomy*, Groningen, Wolters-Noordhof, 1970.

Baars, J.W.M., *Technology of large Radio Telescopes for millimeter and submillimeter wavelengths*, Infrared and Millimeter Waves, **9**, 241-281, K.J. Button, Ed., New York, Academic Press, 1983.

Baars, J.W.M., B.G. Hooghoudt, P.G. Mezger and M. J. de Jonge, The IRAM 30-m millimeter radio telescope on Pico Veleta, Spain, *Astron. Astrophys.* **175**, 319-326, 1987 [Gold, 56].

Baars, J.W.M., A. Greve, H. Hein, D. Morris, J. Penalver and C. Thum, Design parameters and measured performance of the IRAM 30-m millimeter radio telescope, *Proc.IEEE* **82**, 687-696, 1994.

Baars, J.W.M., R.N. Martin, J.G. Mangum, J.P. McMullin and W.L. Peters, The Heinrich Hertz Telescope and the Submillimeter Telescope Observatory, *Publ. Astron. Soc. Pacific* **111**, 627-646, 1999.

Baars, J.W.M., R. Lucas, J. Mangum and J.A. Lopez-Perez, Near-field radio holography of large reflector antennas, *IEEE Antennas and Propagation Magazine* **48**, 2006.

Bennett, J.C., A.P. Anderson, P.A. McInnes and A.J.T. Whitaker, Microwave holographic metrology of large reflector antennas, *IEEE Trans. Antennas Propagat.* **AP-24**, 295-303, 1976.

Born, M. and E. Wolf, *Principles of Optics*, 6th Ed. Oxford, Pergamon, 1980.

Chu, T.S., R.W. Wilson, R.W. England, D.A., Gray, and W.E. Legg, The Crawford Hill 7-meter millimeter wave antenna, *Bell System Tech. J.* **57**, 1257-1288, 1978.

Conway, R.G., E.J. Daintree and R.J. Long, Observations of radio sources at 612 and 1400 Mc/s, M*onthly Notices Roy. Astron. Soc.* **131**, 159-171, 1965.

Conway, R.G., Measurement of radio sources at centimetre wave-lengths, *Nature* **199**, 1177, 1963.

Cook, J.S., E.M. Elam and H. Zucker, The open Cassegrain antenna, *Bell System Tech. J.* **44**, 1255-1300, 1965.

Davis, M.M., A 1417-MHz search for radio sources having a flux excess at short wavelengths, *Bull. Astron. Institutes of the Netherlands* **19**, 201-226, 1967.

Dicke, R.H., The measurement of thermal radiation at microwave frequencies, R*ev. Sci. Instr.* **17**, 268-275, 1946.

Emerson, D.T., U. Klein and C.G.T. Haslam, A multiple beam technique for overcoming atmospheric limitations to single-dish observations of extended sources, *Astron. Astrophys.* **76**, 92-105, 1979 [Gold, 342].

Findlay, J.W., *Antennas and Receivers for Radio Astronomy*, Advances in Radio Research, Ed. J.A. Saxton, **2**, 37-119, New York, Academic, 1964.

Fried, D.L. and J.D. Cloud, Propagation of an Infinite Plane Wave in a Randomly Inhomogeneous Medium, *J. Opt. Soc. Am*, **56**, 1667-1176, 1966.

Gibson, J., W.J. Welch and Imke de Pater, Accurate jovian radio flux density measurements show ammonia to be subsaturated in the upper troposphere, *Icarus* **173**, 439-446, 2005.

Godwin, M.P., E.P. Schoessow and B.H. Grahl, Improvement of the Effelsberg 100 meter telescope based on holographic reflector surface measurement, *Astron. Astrophys.* **167**, 390-394, 1986 [Gold, 215].

Greve, A., Reflector surface measurement of the IRAM 30-m radio telescope, *Int. J. Infrared and Millimeter Waves* **7**, 121-135, 1986.

Hachenberg, O., B.H. Grahl and R. Wielebinski, The 100-meter radio telescope at Effelsberg, *Proc. IEEE* **61**, 1288-1295, 1973 [Gold, 32].

Hartmann, D., P.M.W. Kalberla, W.B. Burton and U. Mebold, Stray radiation correction as applied to the Leiden/Dwingeloo survey of HI in the Galaxy, *Astron. Astrophys. Suppl. Ser.* **119**, 115-151, 1996.

Hartsuijker, A.P., J.W.M. Baars, S. Drenth and L. Gelato-Volders, Interferometric measurement at 1415 MHz of radiation pattern of paraboloidal antenna at Dwingeloo Radio Observatory, *IEEE Trans.Antennas Propagat.* **AP-20**, 166-176, 1972.

Herring, T.A., Modelling atmospheric delays in the analysis of space geodetic data, *Publ. in Geodesy, Neth. Geodetic Commission, New series* **36**, 157-164, 1992.

Hills, R. et al., High-Resolution Millimetre-Wave Holography on the James Clerk Maxwell Telescope, *Proceedings URSI Assembly*, Maastricht, 2002

Jennison, R.C., *Introduction to Radio Astronomy*, London, Newnes, 1966.

Kildal, P.-S., E. Olsen and J.A. Aas, Losses, sidelobes, and cross polarization caused by feed-support struts in reflector antennas: Design curves, *IEEE Trans. Antennas and Propag.* **36**, 182-190, 1988.

Kolmogorov, A.N., The local structure of turbulence in incompressible viscous fluid for very large Reynolds numbers, *Roy. Soc. Proc., Series A* **434**, 9-13, 1991; translation of paper in Russian, 1941.

Kramers, H.A., La diffusion de la lumière par les atomes, *Atti. cong. int. fisici, Como* **2**, 545-557, 1927. (Also in *Collected Scientific Papers* (North-Holland, Amsterdam), 333–345,1956.)

Kronig, R., On the theory of dispersion of x-rays, *J. Optical Soc. Am.* **12**, 547-557, 1926.

Landau, L.D. and E.M. Lifshitz, *Electrodynamics of continuous media*, Course of theoretical physics, **8**, §62, 256, Oxford, Pergamon, 1960.

Liebe, H.J., MPM—An atmospheric millimeter-wave propagation model, *International Journal of Infrared and Millimeter Waves* **10**, 631-650, 1989.

Love, A.W. (Ed.), *Reflector Antennas*, New York, IEEE press, 1978.

Mangum, J.G., J.W.M. Baars, A. Greve, R. Lucas, R.C. Snel, P.Wallace and M. Holdaway, Evaluation of the ALMA prototype antennas, *Publ. Astron. Soc. Pacific* **118**, 1257-1301, 2006.

Misell, D.L., A method for the solution of the phase problem in electron microscopy, *J. Appl.Phys. D* **6**, 16-19, 1973.

Morris, D., Chromatism in radio telescopes due to blocking and feed scattering, *Astron. Astrophys.* **67**, 221-228, 1978.

Morris, D., Phase retrieval in the radio holography of reflector antennas and radio telescopes, *IEEE Trans. Antennas Propagat.* **AP-33**, 749-755, 1985 [Gold, 220].

Morris, D., J.W.M. Baars, H. Hein, H. Steppe, C. Thum and R. Wohleben, Radio-holographic reflector measurement of the 30-m millimeter radio telescope at 22 GHz with a cosmic signal source, *Astron. Astrophys.* **203**, 399-406 1988.

Morris, D., H. Hein, H. Steppe and J.W.M. Baars, Phase retrieval radio holography in the Fresnel region - Tests on the 30 M telescope at 86 GHz, I*EE Proc. Part H* **135**, 61-64, 1988.

Nikolic, B., J. Richer and R.E. Hills, Phase Retrieval Measurements of Antenna Surfaces Using Astronomical Sources, *Proc. General Assembly of URSI*, 2002.

Padman, R. and R.E. Hills, VSWR reduction for large millimeter-wave Cassegrain radiotelescopes, *Int. J. Infrared and Millimeter Waves* **12**, 589 - 599, 1991.

Pardo, J.R., J. Cernicharo and E. Serabyn, Atmospheric transmission at microwaves (ATM): an improved model for millimeter/submillimeter applications, *IEEE Trans. Antennas and Propag.* **49**, 1683-1694, 2001.

Penzias, A.A. and R.W. Wilson, A Measurement of Excess Antenna Temperature at 4080 Mc/s, *Astrophys. J.* **142**, 419-421, 1965.

Rayleigh, Lord, On pin-hole Photography, *Phil. Magazine* **31**, 87-99, 1891. (*Scientific Papers*, Vol. 3, 429-440, New York, Dover, 1964)

Rudge, A.W. and N.A. Adatia, Offset-parabolic-reflector antennas: A review, *Proc. IEEE* **66**, 1592-1618, 1978

Rüeger, J.M., *Electronic Distance Measurement - An introduction*, Berlin, Springer Verlag, 1996.

Rüeger, J.M., Refractive index formulae for radio waves, *Proc. FIG XXII Int. Congress*, Session JS28, 2002.

Rusch, W.V.T., O. Sorensen and J.W.M. Baars, Radiation cones from feed-support struts of symmetric paraboloidal antennas, *IEEE Trans. Antennas Propag.* **AP-30**, 786-790, 1982.

Schelkunoff, S.A., *Electromagnetic Waves*, 363-365, New York, van Nostrand, 1943.

Scott, P.F. and M. Ryle, A rapid method for measuring the figure of a radio telescope reflector, *Monthly Notices Roy. Astron. Soc.* **178**, 539-545, 1977 [Gold, 200] .

Serabyn, E., T.G. Phillips and C.R. Masson, Surface figure measurements of radio telescopes with a shearing interferometer, *Appl. Optics* **30**, 1227-1241, 1991.

Smart, S., *Textbook on Spherical Astronomy*, 55-60, Cambridge, University Press, 1962

Smith, E.K. and S. Weintraub, The constants in the equation for atmospheric refractive index at radio frequencies, *Proc. IRE* **41**, 1035-1037, 1953.

Stark, A.A., C.F. Gammie, R.W. Wilson, J. Bally, R.A. Linke, C. Heiles and M. Hurwitz, The Bell Laboratories H I survey, *Astrophys. J. Sup.* **79**, 77-104, 1992.

Tatarski, V.I., *Wave propagation in a turbulent medium*, New York, McGraw-Hill, 1961.

Thompson, A.R., J.M. Moran and G.W. Swenson, *Interferometry and Synthesis in Radio Astronomy*, 2nd Ed., New York, John Wiley, 2001.

Tiuri, M., Radio astronomy receivers, *IEEE Trans. Antennas Propagat.* **AP-12**, 930-938, 1964.

von Hoerner, S., Design of large steerable antennas, *Astron. J.* **72**, 35-47, 1967 [Gold, 64].

van Woerden, H., *De Neutrale Waterstof in Orion*, Ph.D. Thesis (in Dutch), Univ. of Groningen, 1962.

Weinreb, S., Effect of Feed-paraboloid Reflections upon Line Receiver Performance on the 140' Telescope, *NRAO Technical Note* (unnumbered), Nov. 1967.

Welch, W. J. et 36 alii, The Berkeley-Illinois-Maryland-Association Millimeter Array, *Publ. Astron. Soc. Pacific* **108**, 93-103, 1996.

Woody, D., E. Serabyn and A. Schinckel, Measurement, Modelling and Adjustment of the 10.4 m Diameter Leighton Telescopes, *Proc. SPIE* **3357**, 474-486, 1998.

7. Design features of some radio telescopes

■ 7.1. Introduction

The first fully steerable parabolic reflectors were second World War radar antennas, developed both in Germany and the allied countries. The developments to ever larger and more accurate antennas has been mainly an activity by radio astronomers and mechanical and structural engineers working with them. The full story of this fascinating development has still to be written. In this chapter we shall not endeavour to undertake such a task.

Rather we plan to present in a concise form the most interesting structural, mechanical and electro-magnetic features of a number of important radio telescopes. The aim is to augment the tools presented in the main body of this book with information on background and realisation of new ideas and technologies in this field. The choice of examples has been guided by two arguments. First, new ideas were introduced in the technology to satisfy the scientific requirements for the instrument. Second, to highlight these original features from the personal viewpoint of the author. This is possible because the author has been privileged to be intimately involved in a number of radio telescope projects, each of which pushed the state of the art in design or performance. It should be emphasised that he has not contributed significantly to the structural and mechanical designs of the selected telescopes. Rather, his role was more concerned with assuring technical solutions in accord with the scientific specifications, managing the design and construction projects and participating in the commissioning of the instruments. Thus we aim in this chapter to feature the inventiveness and craftsmanship of the engineers and technicians in industry, who turned the astronomer's dream into reality under budgetary, logistical and environmental limitations. Often, the original design work has remained unpublished, were it only that new tasks prevented the industrial engineer from properly documenting his earlier achievement. Wherever such publications exist, we shall mention them below, but often the only reference in the open literature is authored by the users of the new instrument, rather than the designers. The sections below do not constitute complete descriptions of the instruments. Rather they highlight features of the designs which the author considers original and contributing to the overall development of the subject of large and accurate reflector antennas.

It is hoped that these summaries may interest the reader who is using the remainder of the book in his work and possibly give him ideas as to improve or change certain features of his own current or future antennas. Where appropriate, we shall draw attention to material from the earlier chapters, as it has been used in the design, construction or commissioning of the radio telescopes to be described.

To recapitulate, the task of a radio telescope or communication antenna is to point in a desired direction with sufficient accuracy and collect the radiation from that direction as efficiently as possible while being insensitive to radiation from other directions. This main characteristic must be preserved if the antenna direction is changed, either to follow the target along its path across the sky or to move from one position to a different one, possibly a large angle away. We require the performance of the antenna to be as little dependent on varying environmental parameters, such as temperature and wind force, as possible. While it is impossible to economically build "ideal" instruments, these requirements are normally considered to be met if the antenna pointing and tracking is accurate and stable to better than one-tenth of the half-power beam width (HPBW) and the reflector surface remains in the prescribed (normally parabolic) shape to better than one-sixteenth of the shortest operational wavelength under all operational conditions. These basic operational specifications are by no means trivial. For instance, a 30-m diameter telescope for millimeter wavelengths, operating up to 250 GHz (1.2 mm wavelength) must have a pointing accuracy of 1 arcsecond and a surface rms error of 75 micrometers. Such accuracies are only achievable with special design methods and judicious choice of materials.

In our choice of telescopes to discuss we use the following guiding principles:

- scientific requirements forcing a shift in the state of the art of telescope design,

- originality of the design methods and material selection,

- success in achieving the scientific and technical goal.

The telescopes to be described below include:

1. Westerbork Synthesis Radio Telescope (WSRT), Netherlands.

2. Max-Planck-Institut für Radioastronomie 100-m telescope at Effelsberg, Germany.

3. 30-m millimeter radio telescopes (MRT) of IRAM near Granada, Spain.

4. Heinrich Hertz Telescope (HHT) of the Arizona Radio Observatory, Tucson, USA.

5. Large Millimeter Telescope (LMT) of the University of Massachusetts and INAOE, Tonantzintla, Mexico.

6. two ALMA prototype antennas, located at the VLA site in New Mexico, USA.

In all these projects, with the exception of the Effelsberg telescope, the author has been closely involved. Our choice leaves several important radio telescopes outside of consideration. In particular, mention should be made of the NRAO 140-ft telescope (see Ch. 1), the 64-m Parkes telescope (Bowen and Minnett, 1963) and the highly successful millimeter and submillimeter antennas of IRAM's Plateau de Bure interferometer (Guilloteau et al, 1992), the James Clerk Maxwell Telescope and the Caltech Submillimeter Observatory, the latter designed and built by Robert Leighton (1978), both located on Hawaii (pictures in Ch. 1). Also the Green Bank Telescope (GBT), shown in Chapter 1, will not be discussed here, despite its many new features. Some key data of the telescopes have been assembled in Table 7.1.

Table 7.1. Data on radio telescopes discussed in this chapter

Telescope	Year	Alti(m)	Dia(m)	D/ϵ	λ_{min} (mm)	Weight(t)
WSRT	1970	20	25	1.7 E4	40	124
Effelsberg	1972	300	100	1.7 E5	3.5	3200
MRT	1985	2850	30	4.2 E5	0.85	400
HHT	1994	3200	10	8.2 E5	0.3	50
LMT	2007	4600	50	6.7 E5	0.85	700
ALMA(proto)	2004	2200	12	7.0 E5	0.35	108 / 80
GBT	2002	800	100	2.0 E5	3.0	7300

■ 7.2. The homologous design method

Until the mid nineteen-sixties the structural design of antennas was based on the principle of providing sufficient stiffness against gravity and wind forces to maintain the required shape of the reflector in all attitudes and under environmental influences. With increasing size and simultaneous need of better surface accuracy this became impractical both in economical and technical terms. Perhaps the best example of the futility of pushing brute-force design methods to the extreme is the effort by the US Navy to build a steerable dish of 180 m diameter to eavesdrop on Russian communications via reflections off the Moon. After years of planning and design the project was stopped without realising more than a huge concrete foundation ring. Parallel to this activity, an astrophysicist at the NRAO in Green Bank, West Virginia, by the name of Sebastian von Hoerner was looking into the possibility of building a 200 m diameter reflector for lunar occultation observations of extragalactic radio sources. Not knowing how to design a steel structure he went from first principles and developed the idea of homologous deformations. He coined this name for a design method in which a telescope structure would be allowed to significantly deform under the elevation-angle dependent influence of gravity, but the deformations would occur in such a way that the resulting reflector surface would retain a parabolic shape. One could in principle keep the surface of a deforming structure in the original parabolic shape by adjusting the position of the surface panels by motorised adjusters. This requires a large amount of adjusters and some way to know the actual deformation of the surface, either by measurement or calculation. But with a homologous behaviour of the reflector only the focal point of the best-fit reflector surface will change in position with varying elevation angle of the antenna. Thus the focal length and direction of the beam axis will have to be determined and the receiver position (or secondary reflector in a Cassegrain system) must track the changes in the focal point. This is much simpler than adjusting the primary surface and, moreover, the required measurement can readily be performed with the aid of strong radio sources (see Chapter 5). Von Hoerner (1967a, 1967b) developed the mathematical basis and a practical iterative method for the design of homologous structures. Because homology is used in just about every new antenna design, we summarise its features here.

In his studies von Hoerner first established a number of "natural limits" to the size of a reflector with a certain surface quality. First there is the stress limit, where the structure would collapse under its own weight. For a steel structure this is about 600 m. A more important limit is the gravitational deformation as function of elevation angle. This deformation is proportional to the length times the mass of the structural member divided by the stiffness. The cross section, i.e. the mass per unit length, typically is proportional to the length and hence the deformation is proportional to the square of the reflector diameter:

$$\delta_g \propto (\rho / E)\, D^2, \tag{7.1}$$

where ρ is the density and E the modulus of elasticity of the material. If we define the shortest wavelength λ_g to be equal to 16 times the rms reflector deformation (see Ch. 4.6), and put in the numbers for the material constants, this delivers the following relation

$$\lambda_g \approx 70\,(D / 100)^2, \tag{7.2}$$

where λ_g is in millimeters and D in meters. Thus in a classical design, based on stiffness, we cannot expect to use a 100 m diameter reflector effectively at wavelengths below 7 cm.

A third natural limit is posed by temperature differences through the structure. The thermal deformation δ_t is proportional to the length of the structural member, its coefficient of thermal expansion C_t and the temperature difference ΔT with respect to its surroundings:

$$\delta_t \propto C_t\, D\, \Delta \mathrm{T}. \tag{7.3}$$

For the minimum wavelength λ_t related to temperature differences we find for a steel structure

$$\lambda_t \approx 6\, \Delta \mathrm{T}\, (D / 100) \tag{7.4}$$

where λ_t is in millimeters, D in meters and ΔT in Kelvin. For aluminium the value is about twice as large. Following von Hoerner we can construct a plot with these limits drawn in and augmented with the actual data from existing antennas. In Fig. 7.1 the ordinate is the "accuracy", defined as the quotient of the diameter D and the surface accuracy ϵ, the abscissa is the diameter. Gravitational and thermal limits are shown, as well as lines of minimum wavelength. There are two groups of antennas: the older ones, designed on the basis of stiffness, all lie under the limits, while the newer group lies above one or both of the limits. Clearly, special tricks must have been used in the design of these antennas to reach their position in the plot. To pass the gravitational limit a homologous design was applied. Beating the thermal limit required temperature control and/or use of special low-expansion materials, like carbon-fiber reinforced plastic (CFRP). Some of these new designs will be described later in this chapter.

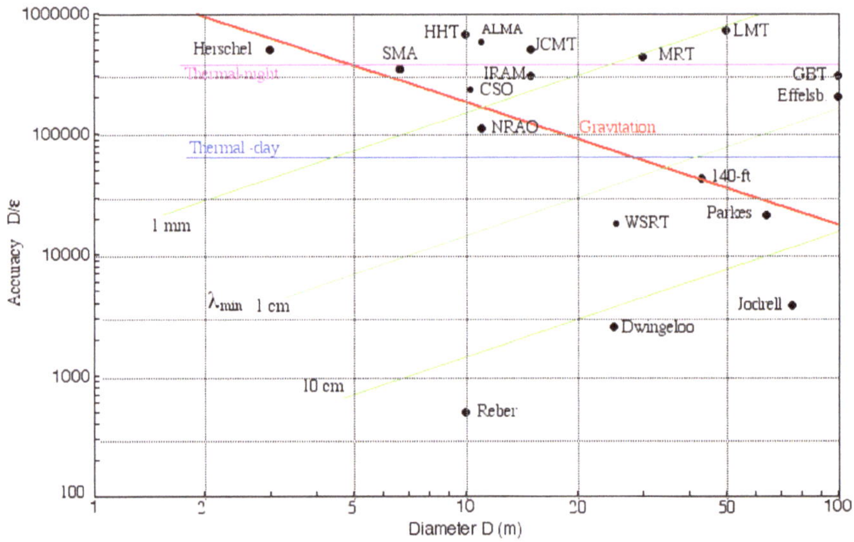

Fig. 7.1. Natural limits in an accuracy versus diameter plot. The positions of millimeter wavelength radio telescopes appear at the top. Older telescopes are below the gravitation limit (red). Most of the antennas have been shown in Ch. 1 or are discussed in this chapter.

A homologous structure must provide an "equal softness" support for the reflector surface. The surface is normally composed of a set of panels and hence the support points of the panels on the backup structure must obey the equal softness requirement. The structural design consists of finding a structure to carry the loads from these points on the surface to the two "hard" points at the elevation bearings in such a way that the surface remains parabolic over the entire range of elevation angles. Von Hoerner proved the existence of a solution and the challenge for the structural designer is to find a practical solution with due regard for the limitations imposed by the geometry (for instance, no two members can cross), the availability of suitably dimensioned structural members (pipes come in standard cross-sections) and weight (which means cost). As we shall indicate below, different approaches can lead to satisfactory results. But homology will never be totally perfect and it is useful to introduce a quantity "deviation from homology" H_0. In an alt-azimuth mounted antenna the gravitational deformations at any elevation angle are described fully by the deformations in zenith and horizon position. Let us call H_z the residual rms deviation between the real and the best-fit paraboloid in zenith position and H_h the similar quantity in horizon position. We can now define the "homology quality", or the deviation from homology, as

$$H_0 = \sqrt{0.5\ (H_z^{\ 2} + H_h^{\ 2})}\,. \tag{7.5}$$

In general, we are interested to minimise the deviations from the best paraboloid in the range of elevation angle where the antenna is predominantly used. For a satellite ground station this is the elevation angle of the satellite. A radio telescope will spend most of the time in the intermediate range of elevation, where the celestial sphere is large and the atmospheric absorption small. Thus there is an incentive to adjust the surface panels to a best paraboloid at an intermediate elevation angle ϕ_0. In that case the deviation from homology, that is the gravitational deformation, as function of elevation angle ϕ can be written as (von Hoerner and Wong, 1975)

$$H_\phi = \sqrt{H_z^{\ 2}\ (\sin\phi - \sin\phi_0)^2 + H_h^{\ 2}\ (\cos\phi - \cos\phi_0)^2}\ . \tag{7.6}$$

The parameters H_z and H_h can be derived from a measurement of the aperture efficiency as function of the elevation angle. Normally they are also obtained from the structural finite element analysis and the comparison with the measurement provides a nice check on the reliability of the calculation. Clearly, if the horizon and zenith deviations are equal, and the whole elevation angle range is considered, the best "rigging angle", the elevation angle where the panels are adjusted to the best fitting paraboloid, will be at 45 degrees elevation. The actual deviations in the

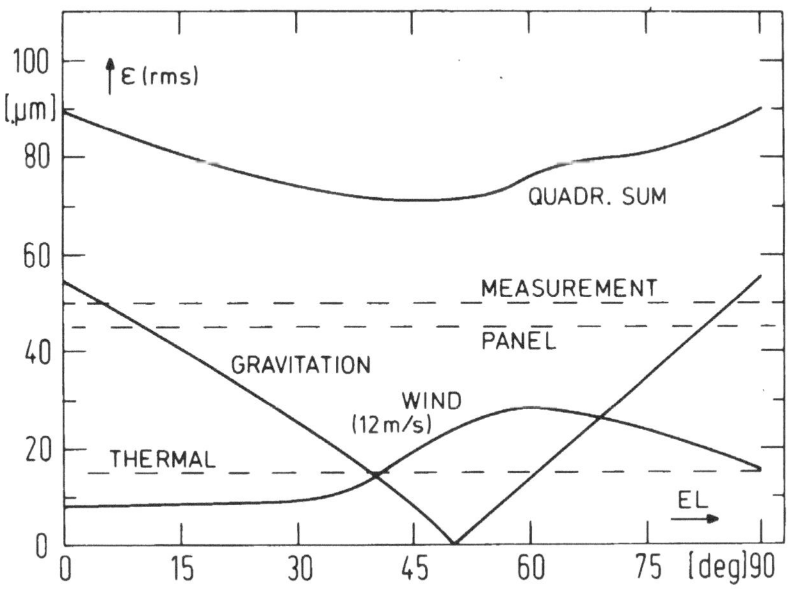

Fig. 7.2. Calculated gravitational and wind induced deformation of the IRAM 30-m millimeter telescope. The wind maximum is offset by the rigging at 50° and the combined error remains below 40 μm between elevation 15 and 75 degrees.

extreme positions zenith or horizon will then be reduced by 25 percent and decrease to zero at the rigging angle. For different zenith and horizon deviations a slightly different rigging angle will be found, while one can also optimise the rigging angle for a limited observation range, for instance between 20 and 80 degrees elevation. Under all circumstances the optimal rigging angle lies between 40 and 50 degrees elevation (von Hoerner and Wong, 1975).

In the context of the optimum rigging angle it is interesting to consider the deformation of the reflector under the influence of wind. It is found that the forces and moments exerted by the wind cause the largest deformations when the horizontal wind impinges on the reflector from the front and the elevation angle is approximately 50 degrees. Setting the surface to the best possible shape near that angle will "smooth" the wind influence on the overall surface error. An example is shown in Fig. 7.2 from the design calculation of the 30-m MRT in Spain. The calculated deviation of homology is 80 μm at 0 and 60 μm at 90 degrees elevation angle, respectively. Aiming at equal surface deformation at elevation angles of 20 and 80 degrees leads to a rigging angle of 50 degrees; the resulting gravitational deformation is the curve dipping to zero at 50°. The wind deformation increases strongly above elevation 30° and reaches a maximum between 50 and 60 degrees. Other error contributions, as temperature, panel fabrication and setting are shown as horizontal dashed lines.The upper curve is the quadratic sum of all deformation components and indicates the overall surface accuracy under 12 m/s frontal wind.

■ 7.3. The Westerbork Synthesis Radio Telescope (WSRT)

Around 1960 Dutch and Belgian astronomers presented a proposal for a large radio telescope in the form of a cross: the Benelux Cross Antenna Project (Christiansen et al., 1963). Development of the "earth rotation synthesis" concept (Ryle, 1962) and limitations in funding, along with the withdrawal of Belgium, led the Dutch to change the proposed instrument to a linear super synthesis array of 12 elements of 25-m diameter antennas (Hooghoudt, 1964). Based on the emphasis in Dutch radio astronomy, established by the work with the Dwingeloo telescope, the original scientific requirement was limited in scope but demanding in quality. The primary goal was to map galaxies with an angular resolution of about 20 arcseconds with high sensitivity, excellent fidelity and full polarisation. A primary observing wavelength of 21 cm was selected, which would also enable the observation of the spectral line of neutral hydrogen at 1420 MHz, once a suitable spectrometer was added to the backend electronics. This led to an array of ten fixed and two moveable elements (on a 300 m long rail track) and a baseline length of 1.6 km (Fig. 7.3), later extended with two more movable elements to 3 km baseline. The quality of the imaging of an interferometric array is critically dependent on the control of the phase of the interferogram during the long integration time. Any differences in the geometry and differential deformation between the element antennas must be minimized if one wants to avoid repeated baseline calibration, or loss of mapping fidelity. For the WSRT it was required that all 12 antennas would be identical in their critical dimensions to 1 mm. The design of the element antennas was led by B.G.Hooghoudt, who had already been heavily involved in the original Dwingeloo telescope design. He also acted as manager of the entire mechanical, electrical and geodetic part of the Project, and from 1966 onward the author was his assistant.

Fig. 7.3. The westerly 5 elements of the WSRT. The fixed antennas are 144 m apart for a total baseline of 1.6 km. The assembly hall is visible in the background. (ASTRON)

As is obvious from Fig. 7.4, the antennas are equatorially mounted. In the mid nineteen sixties this had become an unusual choice for structures of this size, but strongly suggested by the nature of the basic telescope operation. The standard observation would consist of long, mostly 12 hour, source tracking during which the orientation of the east-west baseline rotates with respect to the sky, thereby providing the north-south component in the synthesized aperture plane. With an equatorially mounted antenna this means a single motion about the polar axis with the constant speed of the earth's rotation. In the declination coordinate only the correct value needs to be set before the start of the observation. A simpler positioning and control system is hardly conceivable. And this was precisely what the original design entailed. When a future commissioner explained that he would not be able to determine pointing corrections or obtain beam maps with such a limited control system, a scanning mode at relatively low speed was added to the control system of both axes. This example illustrates the importance of close cooperation of contractor design engineers, project management and future users in the design process of a new telescope.

The choice of an equatorial mount is also highly beneficial to the polarisation capability of the instrument. Now the orientation of the instrumental polarisation plane does not rotate with respect to the source during the long observation, avoiding any variation of instrumental polarisation in the observed signal. These were good reasons to select an equatorial mount despite the fact that normally they tend to be more expensive than the usual azimuthal mounting. Considering the need for 12 identical antennas, the designers were placed for some challenging requirements: design and built a set of 12, highly identical and stable equatorial antennas of 25 m diameter, optimized for 20 cm wavelength, at the lowest possible price.

From Fig. 7.4 the antennas appear as very simple structures and indeed they are. The antenna can be divided into a small number of major parts, shown in the exploded view of Fig. 7.5 (Baars and Hooghoudt, 1974). These are the sections hoisted into place on the foundations along the baseline. The sections themselves are

Fig. 7.4. Element antenna of the WSRT with an equatorial mount, a 25 m diameter wire-mesh reflector and primary focus receiver box. After setting up the base pentapod, the assembly required only 4 lifting maneuvers to complete the antenna, (Author's photo)

assembled in a large, temperature controlled hall at the telescope site, whereby extensive use is made of templates. This assures identical products for all 12 antennas and eases the assembly of the large sections. Basic parts of the sections were manufactured in the shop to the largest size suitable for transportation. For instance the large circular gear racks of the axis movements were transported by ship to about 10 km from the telescope site. Overall, the design aimed at easy reproduction with high accuracy and minimizing on-site activity. The polar axis house was welded in the controlled environment of the hall with the aid of a template. This assured a

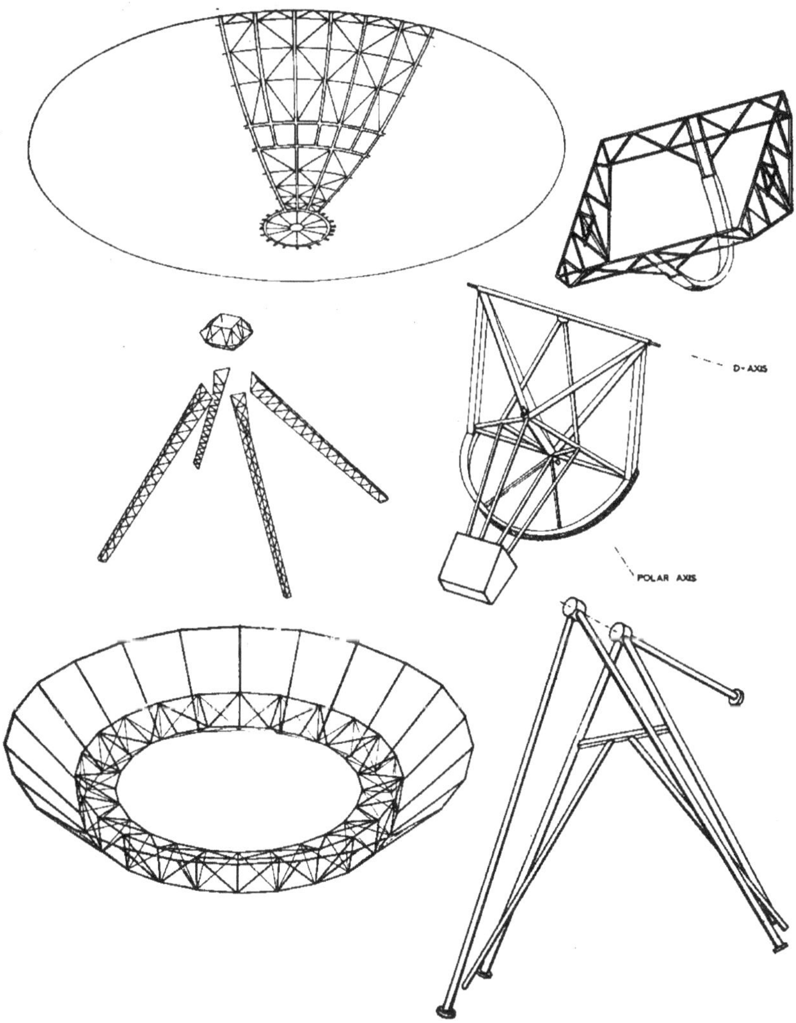

Fig. 7.5. The major section of the antenna in an exploded view. After hoisting the polar-axis house onto the pentapod (lower right), the declination cradle was attached (upper right), followed by the polar counterweight (lower part middle right). The ring girder and reflector surface were assembled in a template and, together with the quadrupod (middle right), lifted in one piece onto the declination cradle.

constant distance and orientation between the polar and declination axes; an accuracy of ±0.2 mm and ±3 arcseconds was achieved for these, respectively.

The declination cradle provides a four-point support for the ring girder, which carries the reflector surface. The four-point support reduces the flexure of the reflector surface by about an order of magnitude with respect to the traditional two-point support. With some imagination one could say that this is a first step towards a homologous design of "equal softness" support for the reflector. The reflector ring-girder arrived in four sections from the shop and was welded together while placed on four pillars, representing the corners of the declination cradle, inside the template for the reflector assembly (Fig. 7.6). This template provides support points for the reflector panels, which were set to within 0.2 mm from the prescribed parabolic form. While placed in the template the 98 panels, divided over three rings, are bolted together to form the reflector surface. The frames of the panels form the radial and tangential members of the reflector. The inner section of the surface is a shell-like membrane structure supported by the ring-girder at its inner radius only. The outer panel ring receives additional support at about 80 percent of the radius by a hoop cantilevered from the ring girder. For the connection of the reflector to the ring-girder a novel solution was introduced. At each connection point a pin from the panel is protruding into a socket on the ring-girder. The socket is filled with epoxy resin to realise a fixed connection. This arrangement allows for normal manufacturing tolerances of the steel structures and the use of epoxy avoids any stress to be exerted on the reflector surface while connecting it to the support ring, thereby maintaining the intrinsic high accuracy of the template assembly.

The reflector surface is a stainless steel mesh of 8 mm spacing, 0.8 mm wire thickness, epoxy bonded to the panel frames. Despite extreme care of handling during manufacture, these bonds delaminated after about ten years. In a repair program, in which many Dutch radio astronomers participated, the mesh surfaces were then riveted to the frames in the field to maintain the correct shape. The epoxy bonding of the main structural parts has not shown any sign of deterioration after more than 35 years. A "pentapod" of five tubular members supports the antenna on the three concrete foundation pads. Because of the relatively high latitude of the site (50 degrees north), a single southern support beam could be placed along the polar axis without affecting the southern sky coverage. This provides for an optimum load transfer to the ground and also gives the antenna a sleek look. The assembly of the antenna in the field required only four hoisting maneuvers and the attachment of a small number of bolts. The antennas were erected with 5 weeks intervals between August 1967 and November 1968.

The application of templates in fabrication and assembly, along with the use of an all-weather assembly hall, were essential ingredients in achieving a highly identical product series. An extensive program of measurements accompanied the construction of the antennas. All these aspects contributed to a radio telescope of superb sensitivity and stability. The reflectors exhibit an rms surface accuracy of 1.4 mm, providing excellent performance at 6 cm wavelength (see Table 5.2). As a result, development of 6 cm receivers was started even before the full array had been completed, increasing the angular resolution by more than a factor three. The high uniformity and stability of the array provides a very high dynamic range, a necessity for the mapping of extended objects with a large range in brightness. After a few years of successful

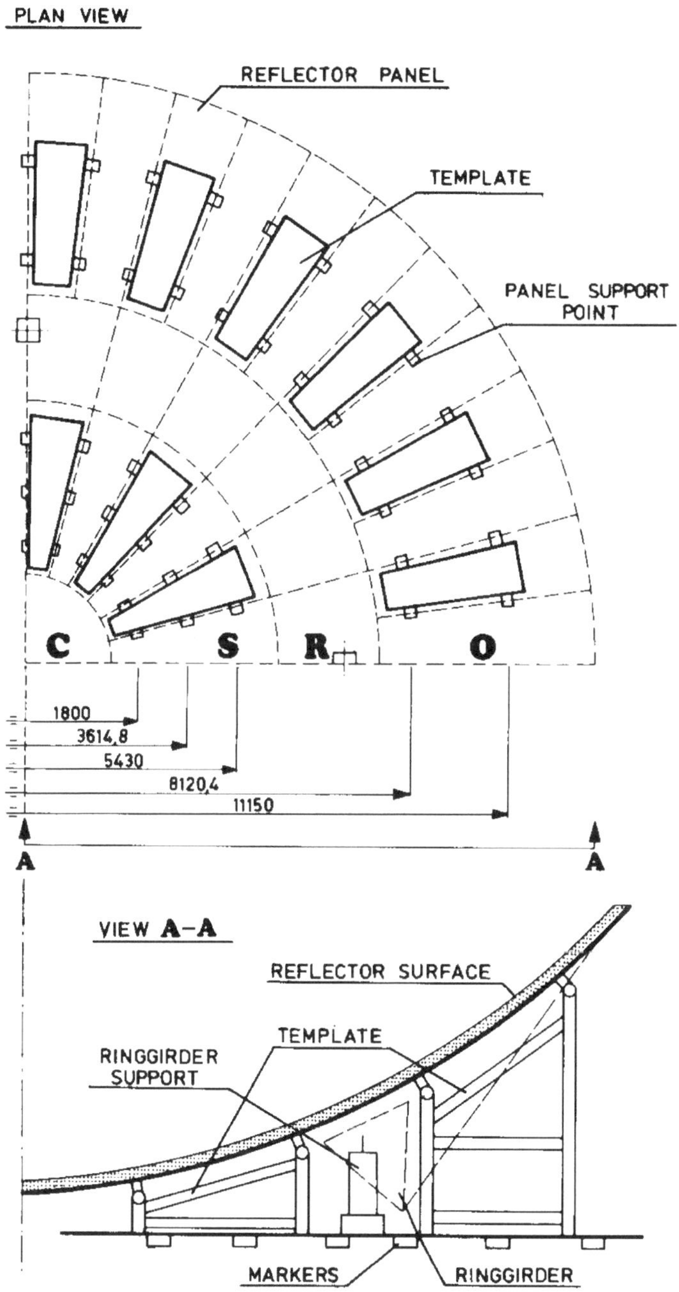

Fig. 7.6. The template for the assembly of the reflector in the assembly hall, consisting of two concentric rings of support beams with accurately located support points for the panels. Before the reflector was assembled the ring girder was placed inside the template, supported on 4 pillars, mimicking the 4 corners of the declination cradle.

operation, the array could be extended with two more antennas, placed one mile to the east, thereby doubling the baseline and increasing the angular resolution by a factor of two. Other extensions and improvements have been made in the area of electronics and data reduction.

After more than 35 years in full operation, the WSRT remains heavily subscribed and continues to play an important role in several areas of radio astronomical research. At the beginning of this successful life stood the need to descope significantly from the original gigantic Benelux Cross Antenna Project to an instrument the Dutch could afford on their own. Earth-rotation synthesis showed the way. The budget situation forced the choices and the choice was made to build a telescope with excellent performance at one wavelength of 21 cm only. No concessions were made in the quality of the most expensive part, the 12 antennas, but economy of scale played an important role in the design process. Making full use of the "mass-production" aspect, introducing novel technologies and exercising a rigorous quality control, the engineers delivered a telescope within budget and schedule which formed the basis for a highly flexible and productive telescope. Currently the WSRT operates at six frequency bands, with capabilities for spectroscopy and pulsar observations and it forms a sensitive station in the European VLBI network.

The choice of an equatorial mount was already unusual in 1965 and has since then been essentially discarded. In the author's opinion this may not necessarily always be the best approach. In particular, it could be worthwhile to consider this choice for an array of many relatively small antennas (10 – 20 m say), which will operate essentially in aperture synthesis mode with long tracking observations. Thus it could be a suitable solution for the elements of the Karoo Array Telescope (KAT) in South Africa and the "many small antennas" version of the Square Kilometer Array (SKA).

■ 7.4. The Effelsberg 100-m radio telescope

Radio astronomy in Germany was established at the University of Bonn with the construction of a 25-m telescope in 1957. In 1962 Otto Hachenberg was appointed professor of radio astronomy. He came from the Heinrich-Hertz-Institute in Berlin, where, together with industry, he had built a 32-m transit instrument. Lacking finite element analysis and computing power, they had studied experimentally how space-frame structures deform under gravity. Here an indication of homologous deformation became apparent without the analytical foundation, provided later by von Hoerner. In Bonn, Hachenberg began to plan for a giant fully steerable telescope and a proposal for financing was presented to the Volkswagen-Foundation in 1964. The reaction was positive under the condition that the capability of an effective operation of the instrument could be demonstrated. This led to the establishment of the Max-Planck-Institute for Radio Astronomy in Bonn in 1965, of which Hachenberg became the founding director.

Design studies for a 80-m diameter antenna were made by Krupp and MAN, where the ideas of homology, as experimentally experienced by Hachenberg and theoretically worked out by von Hoerner, were applied. Eventually this led to a joint venture of the two companies in 1965 with the task of designing and building a fully steerable telescope of 100 m diameter suitable for observation at a smallest wavelength of

2-3 cm. This requires a reflector accuracy of ~1 mm and a pointing accuracy of <10 arcseconds, both a significant improvement over existing large telescopes, all of which were much smaller than 100 m (Hachenberg, 1968, Geldmacher, 1970, Altmann, 1972).

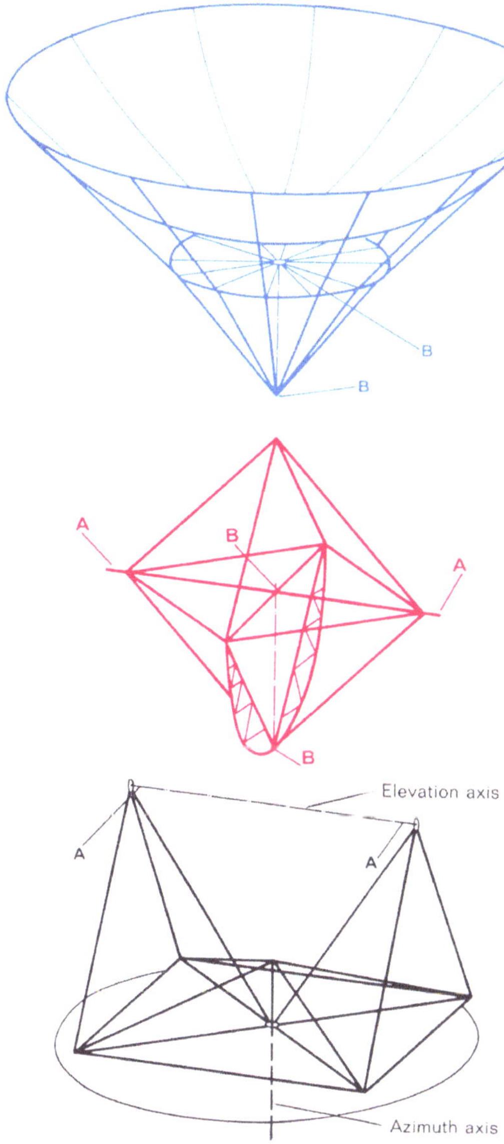

Fig. 7.7. Exploded view of the three main sections of the antenna. The upper section is the BUS, supporting the panel units through a space frame structure. The middle octahedron supports the BUS at points B and is itself supported on the alidade at points A. The upper half of the octahedron forms the quadripod support of the subreflector. (Drawing Krupp)

The reflector backup structure (BUS) was conceived as a spaceframe with a high degree of rotational symmetry. This was guided not only by the idea of homology, but also by the still limited computational power of contemporary computers. For the structural analysis of the BUS only a pie-shaped section of one-twenty-fourth needed to be considered. The engineers achieved the desired result with a method of "trial and success", combining their practical experience with the new ideas of homologous structures. The BUS is supported from the back by an umbrella-type cone of spokes (Fig. 7.7). This structure (blue) is attached to the elevation structure (red) at only two points B: the tip of the umbrella and the center of the spoke wheel in the lower plane of the BUS proper. This two-point suspension provides a stiffness which is symmetrical with respect to rotation. This in turn assures a homologous behaviour in which the

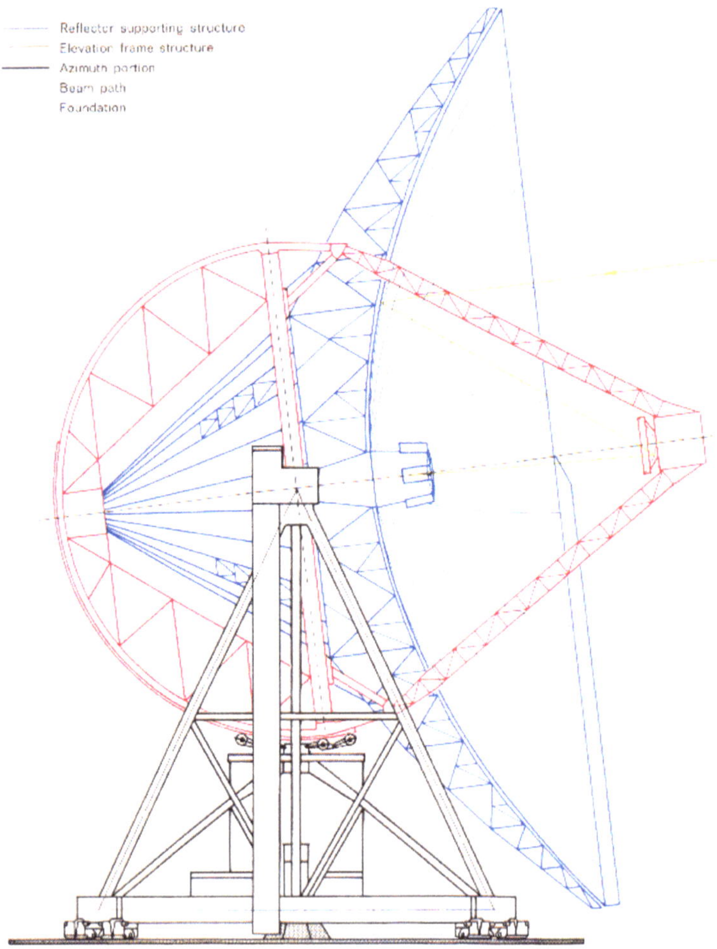

Fig. 7.8. Cross-section through the Effelsberg 100-m telescope. The reflector support structure (blue) is highly symmetrical and is connected to the elevation octahedron (red) at two points only. The octahedron is supported at two elevation bearings by the alidade (black), which runs on a 64 m diameter railtrack. (Drawing Krupp)

elastic deformations under varying elevation angle will result in a surface close to a paraboloid. The elevation structure (red) is essentially an octahedron, as suggested by von Hoerner (1967) in his classic paper. The quadripod support for the primary focus cabin / subreflector forms one half of the octahedron. The lower half is composed of the elevation bull-gear and two beams to the elevation bearings, where the entire structure is supported at points A by the alidade (black). This solution assures that the quadripod does not have any influence on the reflector. Also there is no physical connection between the octahedron near the elevation bearings and the BUS. Contrary to most earlier designs the surface is not subjected to the "point" loads of the quadripod and the support at the elevation bearings.

Fig. 7.9. The 100-m diameter Effelsberg telescope of the MPIfR, Bonn, Germany. The "umbrella" reflector support and the rear half of the octahedron structure are well visible. The cabins at the elevation bearings are 50 m above the ground. The collar around the edge of the reflector serves to reduce pick-up of ground radiation. (N. Junkes, MPIfR)

The alidade runs on a railtrack of 64 m diameter through 4 bogies with 4 wheels each. The elevation drive originally contained two dual anti-backlash drives, but instabilities in the system forced the removal of one drive, fortunately without ever impeding the operation of the telescope. A cross-section through the telescope is shown in Fig 7.8, where the same colours have been used to show the major sections.

The optics of the Effelsberg telescope is somewhat unusual. First, it employs a Gregorian two-reflector system with an elliptical secondary mirror of 6.5 m diameter. The primary focus is accessible through a hole in the secondary reflector to accommodate long wavelength (>20 cm) feeds. Second, the focal ratio of the primary reflector is 0.3, which makes the dish quite deep. To minimise spill-over radiation from the ground, a vertical shroud extends from the perimeter of the reflector. This was especially useful, because the telescope is located in a narrow, deep valley (Fig. 7.9) to shield it from radio interference.

The surface of the reflector over the inner 65 m diameter consisted originally of aluminium sandwich panels with an rms accuracy of 0.25 mm. Similarly as with the WSRT the epoxy bonding between the surface plates and the honeycomb core delaminated after about ten years and all panels were replaced by aluminium plates reinforced with backing ribs (cassette panels). This type was used *ab initio* for the area between 65 and 80 m diameter. To decrease wind loading, the outer area was originally composed of stainless steel wiremesh with 6 mm mesh size. With the replacement of the sandwich panels, the mesh was also replaced with cassette panels, perforated with 7 mm diameter holes to decrease wind loading. Thus the telescope is not effective beyond the inner 80 m for wavelengths smaller than about 2-3 cm.

The performance of the instrument is quite impressive. After setting the surface with the new panels with the aid of satellite holography at a frequency of 12 GHz, the rms surface error is 0.45 mm at the setting angle of 32 degrees. It increases to only 0.7 mm at an elevation angle of 80 degrees. This increase in surface rms should be compared with the actual deformation of the structure, which is 76 mm at the edge of the dish. The pointing accuracy is better than 10 arcseconds with a repeatability over time scales of one hour of 2 arcseconds (Hachenberg et al., 1973). The telescope is routinely used at 3.5 mm for VLBI observations. The only active control necessary for achieving this performance is the adjustment of the subreflector (or primary focus feed) to the focus of the best-fit paraboloid. During the commissioning of the telescope a perfect match was found between the prediction based on structural calculation and the measured adjustment. This is a good demonstration of the high quality of the structural design. Over the range of elevation angle from 15 to 85 degrees the lateral shift in the focus position is about 120 mm, while the axial shift is 15 mm.

Recently, another telescope of comparable size and performance has come into operation; the Green Bank Telescope (GBT) of NRAO (see Ch. 1). Its offset reflector is slightly larger (110 x 100 m) for "political" as well as technical reasons, as explained in Chapter 6.3.2. The large surface elements of the GBT are adjustable through motor-controlled actuators, which enables a constant rms surface error over the entire elevation range, if the necessary corrections are known. The current level of finite element analysis makes this a viable option. This challenge to the Effelsberg telescope is being met by the installation of an active subreflector of 6.5 m diameter, designed and built by MT-Aerospace. It consists of 96 individually adjustable, accurate (10 μm rms) aluminium facets with an overall surface error of less than 80 μm. Because the gravitational and slowly varying thermal deformations of the primary reflector have a rather large scale-length and can be computed reliably with a finite element analysis, the adjustment of the relatively small number of subreflector facets will restore the full accuracy of the dual-reflector Gregorian system. Undoubtedly, the Effelsberg telescope is a splendid piece of engineering, beautifully executed and an impressive example of the power of homologous design.

■ 7.5. The IRAM 30-m Millimeter Radio Telescope (MRT)

In chapter one we mentioned that the development of large reflector antennas has been spurned strongly by the needs of radio astronomers. Interestingly, this is not entirely the case for the spectral range, called millimeter-wave radio astronomy, which we define as the frequency range from 30 – 300 GHz (wavelength 10 to 1 mm). The first reflectors, sufficiently accurate to operate at 3 mm wavelength, were the 4.9 m diameter antenna at the University of Texas (Tolbert et al.,1965) and the 4.6 m diameter dish of the Aerospace Corporation in El Segundo, California (Jacobs and King, 1965). Both instruments were not built primarily for radio astronomy, but were however quickly "taken over" by radio astronomers. Eugene Epstein became “corporate astronomer” at the Aerospace Corporation and made a career out of making millimeter wavelength observations near the beach of the Pacific Ocean. The Texas antenna was moved to the McDonald Observatory in 1967, where it operated until 2000. It has now been relocated to the 4600 m high site of the LMT in Mexico (see Sec.7.7). In 1964 the National Radio Astronomy Observatory (NRAO) proposed to develop a large millimeter telescope and operate it from a high and dry site to avoid the strong absorption from atmospheric water vapour at these wavelengths. Peter Mezger (1964) of NRAO summarized the prospects of astronomy at 3 mm wavelength in a report with the conclusion that a limited number of objects might be observable: "planets, compact HII-regions, a few quasars, perhaps recombination lines of ionised hydrogen". Today, this would not have resulted in funding, but the NSF provided 1 million dollars to NRAO to pursue the project. The result was the 36-ft telescope on Kitt Peak, which started operation in 1968. With the discovery of the spectral line of Carbon-Monoxide (CO) at 115 GHz with this telescope (Wilson et al,1970), millimeter wavelength astronomy established itself as a major branch of radio astronomy. The story of this highly productive instrument has been told vividly by Gordon (2005) in his book "Recollections of 'Tucson Operations'". Short descriptions of these and other mm-antennas are presented in a "Special Issue on Millimeter Wave Antennas and Propagation" of the IEEE Transactions on Antennas and Propagation, July 1970.

In his position as director at the Max-Planck-Institut für Radioastronomie (MPIfR), Peter Mezger concluded in 1972 that observing capabilities needed to be augmented with a large telescope for millimeter wavelengths. Preliminary studies led to a proposal for a reflector of 30 m diameter with a surface accuracy of 0.1 mm, which would marginally enable observations at the 2→1 spectral line of CO at 230 GHz, where the beamwidth would be about 10 arcsecs. A critical aspect of the proposal was the placement of the telescope at a site suitable for observations at 1 mm wavelength. At about the same time French radio astronomers were making plans for an interferometric array for mm-observations, while plans were being developed also in the United Kingdom for a dedicated telescope for wavelengths even shorter than 1 mm – the submillimeter range of the spectrum. In a collaborative effort, the three groups proposed the creation of a joint observatory for these telescopes on a suitable site. The proposal met with support from the respective governments and resulted in the end of 1975 in negotiations between the Max-Planck-Gesellschaft and the Centre National de Recherche Scientifique about the establishment of a joint mm-observatory, later named IRAM. The British proceeded on their own, which resulted in the JCMT on Hawaii in a collaboration with the Netherlands and later with Canada.

The MPIfR obtained main funding for the Millimeter Radio Telescope (MRT) from the Volkswagen Foundation and started serious design effort in the second half of 1975 with the same joint venture of companies, which built the Effelsberg telescope. At this time the author joined the MPIfR and became Project Manager of the MRT jointly with Ben Hooghoudt.

The specifications of the MRT formed a formidable challenge for the designers, who however had the Effelsberg telescope as a good point of departure. A straightforward downscaling of the 100-m antenna could be expected to show the required small surface error of not more than 0.1 mm rms. There were however serious complications with this. First, we needed to account for the severe weather conditions, which could be expected at the high mountain site, planned for the telescope. These could include icing storms and very high wind speeds. Second, we wanted to arrange for a spacious and easily accessible room for the receivers to ease their operation and maintenance and to allow a quick change of system depending on atmospheric conditions. This precluded prime focus operation and a Cassegrain room in a scaled Effelsberg antenna would be both difficult to access and too small. Thus it was clear that the highly symmetrical "umbrella" support of the Effelsberg telescope had to be abandoned. Nevertheless, a highly homologous reflector structure would be required to fulfill the surface specification.

The pointing specification of 1 arcsec also was well beyond what had been realised in antennas of similar size, and moreover, this accuracy had to be maintained in a wind speed of 10 m/s. Experience with other telescopes had taught us that there must be a match between surface accuracy and pointing for a telescope to be effective. Older instruments often showed a better surface than expected, but this could normally not be exploited because of lack of pointing accuracy and stability. Thus the designers were told that pointing and surface were of equal importance and should remain "matched" under all operational conditions. Next to the wind, we determined that variations and gradients of the temperature throughout the structure could easily cause deformations large enough to jeopardise the telescope performance. Controlling the influence of these thus became a major requirement for the design.

The requirements for the MRT originated in the experience with other telescopes, notably the NRAO 140-ft and 36-ft telescopes, as well as early operation of the Effelsberg telescope. For the first time this was all bundled into one comprehensive specification, more in terms of operational performance than of structural tolerances. Consequently, a very close collaboration between the industrial designers and the astronomers/engineers of the MPIfR was established from the early design phase throughout the commissioning of the instrument.

The partners in the industrial joint venture, Krupp and MAN, started with presenting each three conceptual designs. Of those, the one shown in the cross-section of Fig. 7.10 was chosen. It is a compact, "turnstile", alt-azimuth antenna, well suited for operation exposed to the hostile climate of high mountains. All drive and receiver systems are accommodated in a two-storey cabin. The entire structure, apart from the reflector surface, of course, is covered by heat-insulating panels, which themselves can be heated to avoid excessive deposit of ice during icing storms (Fig. 7.11). By 1975 advanced finite element analysis programs, like NASTRAN and STRUDL, were available along with sufficient computing power, so that complete structures

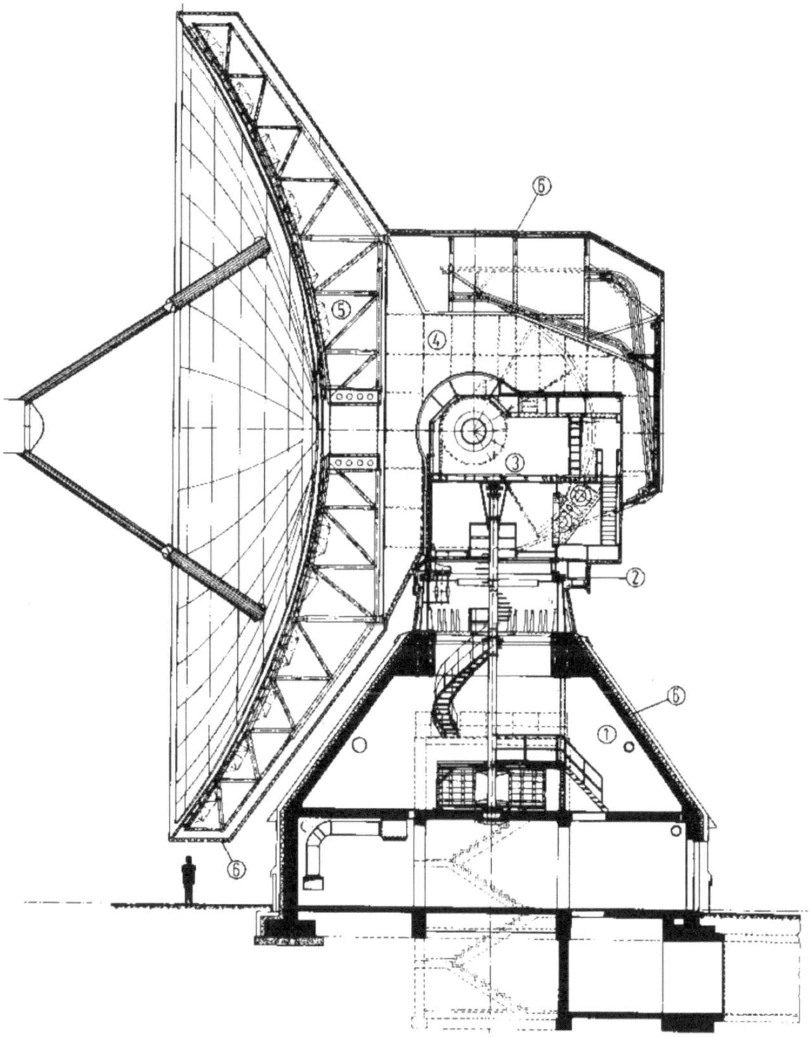

Fig. 7.10. Cross section of the MRT. A concrete pedestal (1) supports a 5 m diameter azimuth bearing (2). The 2-storey cabin (3) is placed between the elevation bearings. A yoke and cone-section (4), supported at the elevation bearings carries the reflector (5). Thermal insulation (6) covers the entire outside of the antenna; it is also present between reflector panels and their support structure.

like the reflector could be analysed. This made the detailed analysis of structures with limited symmetry possible. In the MRT the reflector backup structure (BUS) is composed of a space frame of 20 sections. It is a homologous structure and in order to exploit that, it must be supported by a mount structure which shows equal stiffness at the attachment points. This was realised by a box-like yoke structure, supported at the elevation bearings, with a cone shaped extension terminating in a round, flat plate of 14 m diameter. The BUS is attached to this plate at 40 points on the outer circumference. The entire structure will now behave in a homologous way, if the plate remains flat and round with varying elevation angle. (Brandt and Gatzlaff, 1981, Eschenauer et al., 1980). This structure does not reach a homologous behaviour as well as the Effelsberg design. To obtain the required small deformations a considerable amount of stiffness, hence steel, was put into the structure. Actually, to survive the specified survival loads, 200 km/h wind with 30 cm of ice on the structure, more steel was needed than required for stiffness. This is a general feature of exposed antennas. There was another reason not to concentrate solely on achieving maximum homology. As pointed out above, achieving excellent pointing stability under operational conditions was of equal importance. This led to the idea to incorporate the quadripod support of the subreflector into the overall structural design. Often the

Fig. 7.11. The IRAM 30-m Millimeter Radio Telescope (MRT) at 2850 m on Pico Veleta, Spain. The collar on top is to avoid ice falling from the rim onto the reflector. The reflector itself can be heated from the back to avoid ice deposit during icing storms. (Photo by the author, IRAM)

quadripod has been supported by the mount, independent from the reflector. Here it was made part of the BUS structure and the deformations of the entire structure were optimised to obtain small pointing and reflector errors, both as function of elevation angle and under wind influence. Thus in the MRT the pointing errors caused by the bending of the reflector and the quadripod (see Sec. 5.5.1) compensate each other to a large degree. This is shown in Table 7.2, where the static pointing errors under gravity for the primary and secondary focus are given as function of elevation angle. The pointing jitter under wind of 12 m/s was calculated to be less than one arcsecs.

Table 7.2 Static pointing error due to gravity

Elevation	Static pointing error (arcseconds)	
angle (deg)	primary focus	secondary focus
0	-26.7	8.7
30	-17.5	4.6
50	0.0	0.0
60	11.4	-2.7
90	52.5	-11.5

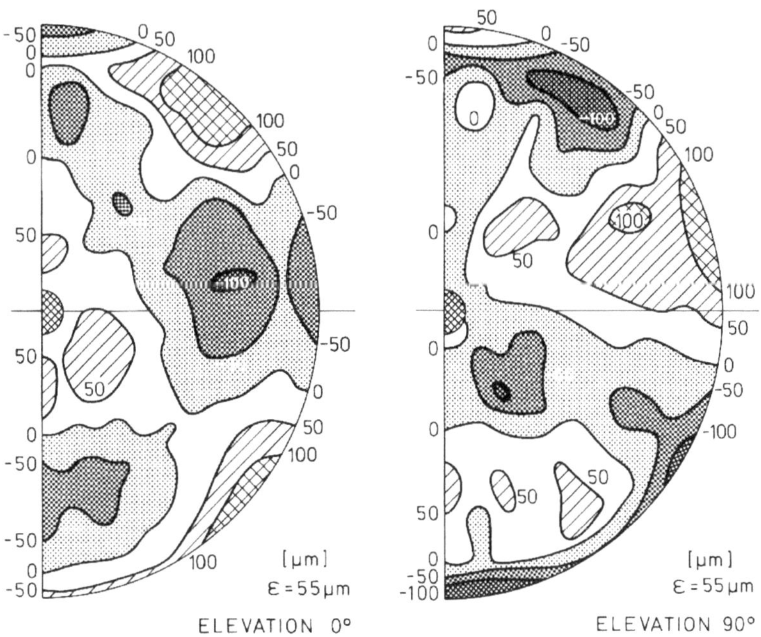

Fig. 7.12. Computed residual deformations (in μm) of the reflector from the best-fit paraboloid in horizon and zenith position, assuming a perfect setting at 50° elevation angle. The rms deformation is 55 μm.

The residual structural deformation as function of elevation, the deviation from perfect homology, is illustrated in Fig. 7.12, where the surface error is shown for horizon and zenith pointing, whereby we assume that the surface was set to the theoretical paraboloid at elevation angle of 50 degrees (See Sec.7.2). This will be successful only if the surface can be measured and hence set at that angle. It was achieved for the MRT by holographic measurement of the surface at 22 GHz with the aid of a strong radio source at the appropriate elevation (Sec.6.1).

There remains to describe the methods used to control the influence of temperature variations. It is clear from a simple calculation, and experienced widely in practice, that the asymmetrical heating of the antenna structure by the Sun can cause large pointing errors and significant deterioration of the surface shape, often in the form of large-scale, comatic or astigmatic deformations. Thus it was clear from the onset that for the MRT these should be kept below 0.1 mm, which means controlling temperature gradients to about 1 K. The first measure was to cover the entire outside of the telescope with insulating panels of polyurethane (Fig. 7.11). Because the site was known to exhibit icing storms, heating elements were embedded in these panels, which are switched on during icing conditions. The concrete pedestal and the massive steel azimuth and elevation structures have a long thermal time constant and maintain a constant temperature to within one degree Celsius during a 24 hour period. Slow seasonal changes are allowed without influencing the performance of the telescope.

The situation is different with the reflector and BUS. During daytime, the Sun will often illuminate (part of) the reflector and despite a thin, white, heat-reflecting paint, a considerable heat flow through the panel must be expected. To keep this heat from entering the BUS volume, insulation was also placed between the panel surface and

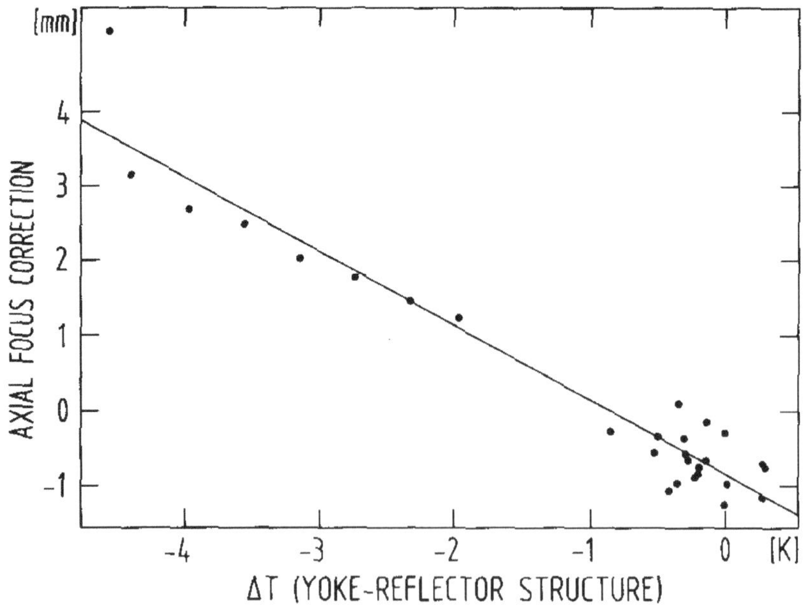

Fig. 7.13. Change in focal length as function of the difference in temperature between BUS and yoke. The stable, linear relationship is routinely used in the focus control software.

the BUS space-frame. The connections from the panels to the BUS are invar, low heat-conducting adjusters. Calculations indicated that the thermal time constant of the BUS was only of the order of one hour and hence a significant daily thermal deformation of the BUS could be expected, despite the insulation around it. We also determined that a difference between the BUS and its supporting yoke-cone structure leads to a symmetrical deformation of the BUS and a concomitant change in focal length. This was later confirmed by measurement and shown in Fig. 7.13. Thus the BUS is kept at the temperature of the yoke by a system of regulated airflow, created by five large fans and a ducting system which creates a slowly circulating airflow to maintain a homogeneous temperature field. The air can be heated or cooled as required. The quadripod is also insulated and maintained at constant temperature by regulated liquid flow through a spiraling tube around the legs.

This system contributes significantly to the stable performance of the antenna. Over the years with long and careful measurements and analysis, some weak points have been identified and improvements were implemented. These are described in an interesting paper by Greve et al (2005). In particular it could be demonstrated that a Finite Element Model of the antenna, loaded with a certain temperature distribution, accurately predicts the resulting deformations. Thus it is now possible to predict by temperature measurements how the surface will deform and which corrections to the panel positions would be needed. Unfortunately, the MRT panels do not have motorised adjusters. But these deformations are normally large scale and correction should be possible by a deformable secondary or tertiary mirror - an example of adaptive optics applied to a radio telescope (Greve et al, 1996).

Summarising, the MRT has surpassed in all aspects the original specifications. For more information see Baars et al., 1987 and 1994. Over the 20 years of its operation the telescope has been improved in performance on the basis of long term measurements and analysis; examples were mentioned above. It is the most powerful instrument for the short millimeter wavelength range (100-350 GHz), but it is being challenged by the 50 m diameter Large Millimeter Telescope (LMT), which we will discuss below.

The MRT uses predominantly the traditional materials steel and aluminium with one exception: the secondary reflector of 2 m diameter employs carbon-fiber reinforced plastic (CFRP). This material was chosen primarily for its low weight, because of the need for a chopping secondary to suppress atmospheric fluctuations (see Sec. 6.3.5). A second great advantage is its very small coefficient of thermal expansion. The mirror was designed and fabricated by Dornier in the form of a composite structure of CFRP skins bonded to an aluminium honeycomb core. The specification of 25 μm could not be guaranteed by Dornier and we agreed to accept the mirror if it was not worse than 50 μm. The fabricator met the challenge magnificently and delivered a secondary with 15 μm rms surface quality. After 20 years of service and more than 5 million chopping cycles, there is no indication of any deterioration in the reflector.

We will now move on to the discussion of a telescope, in which the use of CFRP was unavoidable to satisfy the extreme specifications.

■ 7.6. The Heinrich Hertz Submillimeter Telescope (HHT)

The first good experiences with a CFRP mirror and the excellent thermal and mechanical properties of the material, along with its relative low weight, made it an attractive, be it relatively expensive, choice for structures with specifications more stringent than those of the MRT. The requirements for the mobile antennas of the IRAM Millimeter Interferometer (Guilloteau et al., 1992) included transportability along rail tracks between observing stations, suppression of thermal deformations and a surface accuracy goal of 50 μm. Clearly, the use of CFRP offers an attractive alternative and indeed the material was widely used in the three 15-m diameter antennas. The BUS is a space-frame structure of a combination of steel and CFRP tubular members and the surface panels are a composite of aluminium honeycomb with CFRP skins, similar to the technology used for the MRT subreflector.

This development took place parallel to the construction of the MRT and spurned interest within the MPIfR for the design and construction of an even more accurate telescope, which would be usable to a smallest wavelength of about 300 μm. This submillimeter region of the spectrum was essentially unexplored but promised to be of great interest for astronomy. Two telescopes were under construction in the mid eighties, which aimed to penetrate into this spectral region: the 15-m JCMT and the 10-m Caltech submillimeter telescope, both located on Hawaii (see Ch. 1 for pictures and references).

At the MPIfR we set our goal to design and build a telescope which would complement the MRT towards submillimeter wavelengths. Our goal was to achieve good performance in the wavelength window near 0.3 mm, where the atmosphere can be acceptably transparent from the highest mountain sites. A collaboration with Steward Observatory at the University of Arizona in Tucson was established and the 3200 m high Mt. Graham in the Pinaleno Mountains of eastern Arizona was selected as site.

We started design studies in 1984 with the companies Krupp Antennentechnik and Dornier. It was quickly established that a temperature controlled steel and aluminium structure was not feasible for the stringent requirements on the stability of the antenna. Thus we concentrated on the use of CFRP for major parts of the telescope. In the event, Dornier was replaced by MAN-Technologie in the decisive phases of the design and construction.

The major operational requirements for the submillimeter telescope (SMT) were quite similar to those for the MRT, but several factors contributed to a rather different design. First, we realised early that the much smaller size of the structure would not leave sufficient space for receivers in a cabin behind the vertex of the reflector. The alternative of a Nasmyth focus was chosen, which is realised by placing a flat mirror on the elevation axis under 45° to the beam coming from the secondary reflector, thereby deflecting this beam by 90° along the elevation axis through a hole in the bearing to the outside of the mount. Because only extremely dry, and therefore high, mountain sites will enable observations near 1 THz (and that only for part of the time), it was decided to place the antenna in a protective building, which would be opened during observations. This moves the survival conditions for extreme weather from the antenna to the building.

Next we designed the interior of the building to provide large rooms for receivers directly adjacent to the outside of the elevation bearings, that is around the Nasmyth focus. The building could also easily accommodate a control room and electronic and mechanical workshop. The enclosure, designed at Steward Observatory, is a co-rotating barn. The two roof panels and front doors open wide during observations (Fig. 7.14). With this arrangement, the antenna design need not be concerned with survival conditions and can concentrate on minimising gravitational, wind and thermal deformations under observing conditions. Because of the strong absorption by water vapour at submillimeter wavelengths, these conditions imply a clear sky, but it can include relatively strong wind.

Fig. 7.14. The HHT on Mt. Graham, Arizona. The enclosure rotates with the telescope. Receiver and control rooms are just visible behind the reflector. The doors and roof panels are closed during inclement weather. (Photo by the author)

It was decided from the onset to realise the reflector backup structure (BUS) in CFRP. A major complication in the design of an all-CFRP space-frame structure is presented by the nodes. In the mid nineties (and even at the time of writing, 2006) a node where up to eight or nine tubes are joined poses a formidable design and manufacturing challenge, when realised in CFRP. Thus, the early concepts for the BUS, worked out by Dornier, considered a BUS in the form of a box structure from epoxied and riveted CFRP plates. A viable construction was designed, but not realised. Eventually, the construction contract was placed with Krupp Antennentech-nik for the entire mount and BUS and to MAN Technologie for the reflector panels. Krupp decided to apply their great expertise in space-frame design on a structure made of CFRP. The node problem was summarily solved by using invar steel. This has the interesting consequence that the weight of the steel nodes is about equal to that of the CFRP tubes. Because CFRP is about 4 times lighter than steel, the overall weight of the BUS is still rather small, compared to similar steel structures.

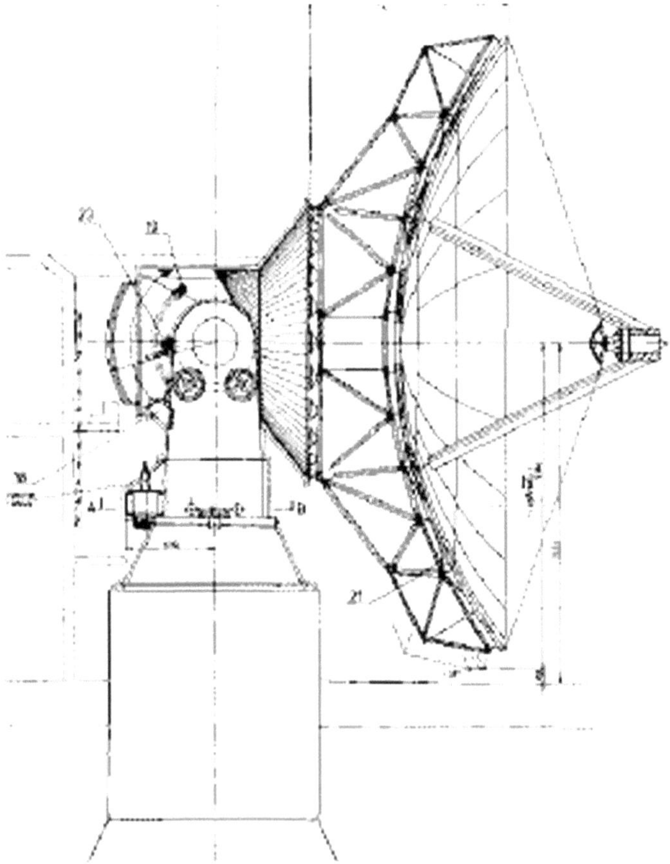

Fig. 7.15. Cross-section through the HHT. The concrete pier carries the antenna; the azimuth and elevation sections are made of steel and are insulated to stabilise the temperature. The BUS and the quadripod are made of CFRP.

The composite nature of CFRP presents the design engineer with additional free parameters in the layout of the structure and the detailed composition of the structural members. The fibers can be obtained with different values for the modulus of elasticity and the percentage of fiber in the final composite CFRP tube can also be varied, as well as the way the fibers are embedded in the resin. Thus stiffness, thermal expansion coefficient and even hygroscopic behaviour can be chosen to obtain an optimal structural behaviour under gravity and varying thermal and humidity conditions. The design group at Krupp undertook the challenge to perform a simultaneous multi-parameter optimisation of the CFRP members and space-frame structure as a whole with the goal to obtain a structure with an optimum distribution of gravitational, thermal and humidity deformations (Stenvers and Wilms, 1989, Mäder et al., 1990). This design effort was accompanied by laboratory experiments to check the calculated parameters against measurements. As an example we show the results on a single CFRP member in Table 7.3.

Table 7.3. Calculated and measured parameters of structural member of CFRP tube with invar steel node

Parameter	Calculation		Measurement	
	Tube	Member	Tube	Member
Compliance (μm kN^{-1})	8.29	10.75	8.16	10.76
Thermal expansion coef. (10^{-6} K^{-1})	0.33	0.62	0.31	0.65
Swelling coef. (10^{-6} $percent^{-1}$)	158	121	$\simeq$120	---

A cross-section drawing of the antenna is shown in Fig. 7.15. The mount is traditional and made of steel. The structural optimisation included the steel section of the elevation structure, the CFRP BUS and the CFRP quadripod simultaneously. The cone-section of the elevation structure terminate in a flat disc and the CFRP BUS is supported by 24 steel blades ("knife-edges") placed along the outer circumference of the disc. The new aspect of this design is the optimisation of the structure for several loading conditions simultaneously. The design team succeeded beautifully and the specifications were significantly surpassed for most conditions.

The fabrication of the reflector panels also involved some new aspects. MAN Technologie had obtained considerable experience in the replication of CFRP composite panels from steel molds for the antennas of the IRAM interferometer. Our specification of 7 μm manufacturing accuracy was however more than twice as severe. The panels are composed of an aluminium honeycomb core to which CFRP plates are bonded on both sides. The parabolic form is obtained by lay-up of the CFRP prepreg on the curved mold and the panel on the mold is cured at high temperature to obtain its final shape. Two new elements enabled us to achieve the specified goal. For the molds we used pyrex glass, ground to 3 μm rms accuracy at the Optical Sciences Center of the University of Arizona. The design of the composite panel aimed at a thermal expansion coefficient of $3 \cdot 10^{-6}$ K^{-1}, equal to that of the mold material. This would minimize the introduction of stresses during the heating and cooling procedures of the fabrication. However, a slight warping of the panel after release from the mold could be expected. These were corrected in the final surface measurement by the existence of 5 or 6 adjusters per panel. Small scale errors within the panel surface were removed by a second lay-up on the mold in which the reflecting aluminium foil of 40 μm thickness was bonded to the panel in a cold-curing process. A consequence of this procedure is that the physical temperature of the panels must stay below 50 C. This is not a problem at the site which is at 3200 m elevation, but it required caution during storage over a Tucson summer where air temperatures of more than 40 C are not uncommon.

The overall performance of the antenna, along with the major specifications, is summarised in Table 7.4. All specifications have been surpassed and the final overall rms surface error is about 12 μm, the lowest value achieved to date with a 10 m diameter antenna. The surface measurement was done with holography, using the LES 8 satellite at a frequency of 37 GHz (Baars et al., 1999).

Table 7.4. HHT specifications and achieved performance

Error component	Specification	Achieved
Homology imperfection, assembly	7	< 3
Space-frame residual deformation	7	< 6
Panel fabrication	7	6
Panel residual deformation	7	< 5
rss of structural/fabrication error	14	10
Reflector setting allowance	10	7
Overall rss error (μm)	17	12

■ 7.7. The Large Millimeter Telescope (LMT)

With the construction of the HHT we had reached a new record in the size of a radio telescope, if one takes the shortest operational wavelength as the unit of length. This is illustrated in Fig. 7.1, where the "precision" of the reflector is plotted against diameter. Taking the minimum wavelength as 16 ϵ, we see that the most accurate telescopes reach a resolution of 5 - 10 arcseconds at the shortest wavelength. In view of the atmospheric absorption at frequencies above 1 THz it seems barely worthwhile to push for increased surface accuracy to 10 μm or less. Even at the best terrestrial sites, now under development, like Chajnantor at 5000 - 5500 m altitude in northern Chile or the South pole, the windows between 1 and 2 THz will "open up" only on rare occasions. But on a good site with long-term adequate transmissivity up to 400 GHz, one might consider operating a telescope with a surface accuracy comparable to that of the MRT, but with a significantly larger diameter.

This idea was developed during 1994 in discussions between astronomers from the "Instituto nacional de Astrofísica, Óptica y Electrónica" (INAOE) in Tonantzintla, Mexico and the University of Massachusetts (UMass) in Amherst, MA, USA. Both were looking towards expansion of their activities. UMass had operated for almost 20 years a 14 m millimeter telescope nearby on a mediocre site and there appeared room for a new large astronomy project in Mexico. The two institutes proposed to jointly build and operate a large mm-telescope on a very high mountain in Mexico.

This resulted in the LMT/GTM-Project (Large Millimeter Telescope / Gran Telescopio Milimétrico) with a goal to design and construct a 50 m diameter Cassegrain antenna on the 4600 m high Cerro la Negra in central Mexico. The scientific desire was to observe to a wavelength of 1 mm, perhaps 0.8 mm, and hence the primary specification was set to 75 μm reflector surface rms error, with a goal of 70 μm, and a pointing accuracy and stability of 1 arcsecond, goal 0.6 arcsecond. These are formidable specifications, which have barely been met with the MRT, which with its diameter of 30 m has just over one third of the reflector area of the LMT.

The first conceptual design for the telescope placed the antenna inside a fixed radome with a skin transparent to millimeter waves. This was a natural way (the 14 m telescope was also enclosed in a radome) to avoid the influences of wind and the extreme weather conditions directly on the antenna. Thus the antenna could be light-weight, because survival conditions are essentially moved to the radome. But there are serious issues with other aspects of the operation. As we have noted before, the thermal equilibrium of the telescope structure is of the utmost importance to reach the specifications. This is difficult to realise in a closed space with transparent walls. Strong stratification of the air will occur, leading to large temperature gradients over the height of the structure, unless the air is mixed by large blowers inside the radome. This would be tantamount to placing the antenna constantly in a rather strong wind. Then there is the need for the radome to survive the extreme conditions of storm and possibly snow and ice. This amounts to putting the material saved for the antenna into the support structure of the radome. Finally, and most severely, any currently available radome fabric with sufficient strength causes significant loss to the received radiation at short mm-wavelengths. For a discussion of these aspects, see Baars (1983).

Eventually, it was decided to consider also an "exposed" design. After all, the MRT in Spain had been operating successfully for about 10 years in extreme conditions similar to those on Cerro la Negra. Thus in September 1997 a request for quotation was issued, which gave bidders the option of a radome enclosed or exposed solution. Three offers were received with highly different characteristics. One was a relatively lightweight antenna in a radome of 60 m diameter, the others were exposed telescopes. A highly optimised homologous structure with a BUS completely realised in CFRP was posed against a homologous steel structure with an "active" reflector surface consisting of 180 sectors each controlled in position by motorised adjusters. The choice went to the last design, the main aspects of which we summarise now.

The LMT design originates with MAN Technologie and shows strong similarities with the 30 m MRT. A cross-section of the telescope is shown in Fig. 7.16. There are however a number of differences and original design features which deserve description. These originate in the awareness that a structure of this size and performance cannot be built passively for an economic price, if at all. It will be necessary to introduce "active" features in the design to satisfy the major specifications of surface and pointing accuracy. In the LMT these are:

- actuator system for the correction of the reflector surface contour,
- sensor system for the main structural deformations, mainly in the mount,
- optical positioning system for the secondary reflector.

Starting from an optimised structural design (homology) these active systems can reduce the rms surface error and improve the pointing accuracy by about a factor 5 under the defined environmental conditions. To reach our goal we have thus added electrical, mechanical and optical elements to the purely structural design of the telescope; this procedure is called *mechatronics* (Kärcher, 1999, 2006).

A structure similar to that of the MRT, but enlarged to 50 m diameter, will show deviations from pure homology about three times as large, because the deformations scale with the square of the diameter. Thus we must count with a surface deviation on the basis of gravity only of the order of 200 μm. The second major problem is posed by the pointing specification. It is unlikely that a 50 m diameter "sail" in the operational wind of 10 m/s can be held stable to one arcsecond without some fast, active correction mechanism. Also, the deformations due to temperature gradients must be controlled or corrected which is twice as hard as in the case of the MRT (see Sec.7.2). These facts force us to employ the mechatronics aspects in the design.

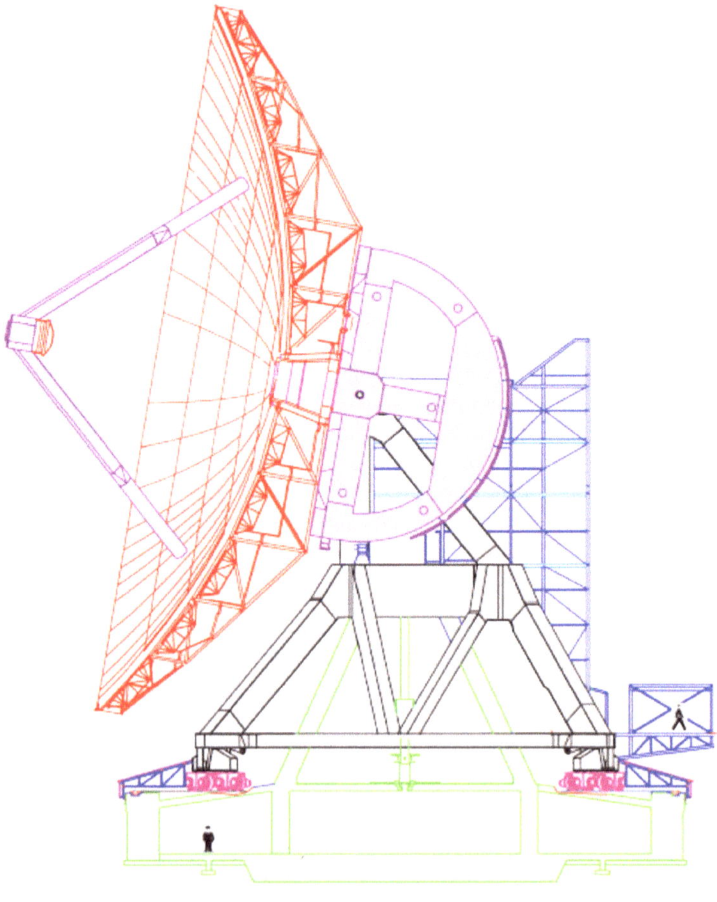

Fig. 7.16. Cross-section through the LMT. The green concrete foundation carries the antenna on 4 bogies and a bearing atop the central concrete pier. The alidade (gray) and elevation section (purple) are made of steel, as are the BUS (brown) and quadripod. Note the man in the basement for scale. (Drawing LMT Project)

But there are more mundane problems in going from 30 to 50 m diameter. For instance, it was determined that the single, large azimuth roller bearing, used in the MRT, could not be realised. In the LMT alidade the vertical load is carried by 4 bogies, each with 4 wheels, onto the rail track with a diameter of 25 m. The lateral loads are carried by a pintle bearing which is placed on top of a central concrete pillar. This minimises turn-over moments and contributes significantly to reducing pointing errors originating in the alidade. In the elevation yoke the support for the reflector backup structure (BUS) is realised by the four corner points of the ballast arms, quite similar to the solution presented before for the WSRT antennas. The BUS is a homologous space-frame structure. The calculated gravitational deformations in zenith and horizon position are illustrated in Fig. 7.17. These are more than a factor ten larger than allowed in the specification and will be compensated by the active surface. The accuracy of current finite element analysis programs is such that this correction can be performed reliably. The stiffness of the BUS is sufficient to keep wind induced deformations of the reflector at an acceptable level without the need to correct these in real time. Such a correction would require a fast, real-time measurement of the deformations or the detailed wind pressure over the surface from which the deformations could be calculated. The 100-m diameter Green Bank Telescope has provisions for such a measurement but it has not yet been fully implemented.

As was mentioned earlier, a serious source of deformation is caused by temperature differences in the structure. The solutions chosen for the LMT are quite similar to those of the MRT. All steel parts are covered by thermal insulation and the air in the BUS volume is constantly circulated by fans to avoid stratification. There remains the problem of the vastly different thermal time constant between the space-frame of

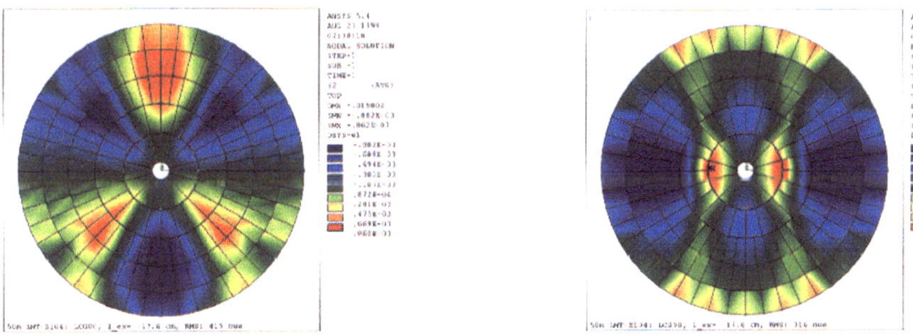

Horizontal position 415 μm rms **Zenith position 316 μm rms**

Fig. 7.17. Computed reflector deformations of the LMT from gravity for the zenith and horizon positions. The values 300-400 μm are so much above the specification that active control of the surface will be necessary.

the BUS and the heavy thick members of the alidade. This necessarily leads to a slowly changing temperature difference between BUS and the alidade and a concomitant large-scale deformation of the BUS. This behaviour is very similar to that found in the MRT. The presence of remotely controlled surface adjustment enables us to correct for these deformations on the basis of calculations using measured temperature fields in the structure. We described this procedure in the discussion of the MRT.

We come now to the important aspect of the pointing accuracy. For a telescope of this size with the best design solutions and using all earlier experience we cannot expect to obtain a pointing stability of better than 3-5 arcseconds in the specified operational wind of 10 m/s. So also here we shall need to introduce active elements in addition to the encoder-servo system. Kärcher (1999) has suggested the term "flexible body compensation" (FBC). The temperature measurements leading to a reflector adjustment, just discussed are an example of such an FBC. For correcting the pointing errors we need sensors for the measurement of those structural deformations which lead to pointing change without being sensed by the angle encoders on the main axes. Those caused by gravity will be constant in time and only dependent on elevation angle. They are encompassed in the overall pointing model (see Sec. 5.5). The errors of a dynamical nature, caused by the wind, lead to bending mainly of the alidade. We remarked already that the reflector itself is sufficiently stiff to avoid significant deformation under wind influence. It acts as a "sail" and causes moments on the alidade leading to pointing errors. In the LMT a set of inclinometers, placed on top of the elevation bearings, senses the bending in elevation and cross-elevation direction. Together with a highly optimised servo-control system this brings the pointing errors within the specification. Simulations indicate that the FBC system reduces the wind induced pointing errors by about an order of magnitude to an rms error of 0.8 arcsecond in 10 m/s wind.

As a last item, the position of the secondary reflector is also monitored by a laser tracker system. This enables corrections for quadripod deformation. This is a rather straightforward extension of the actuator control of the subreflector, needed to keep it positioned in the best-fit focal point of the primary reflector. Position shifts due to temperature and wind influences are sensed by the laser tracker and correction signals are applied to the actuators.

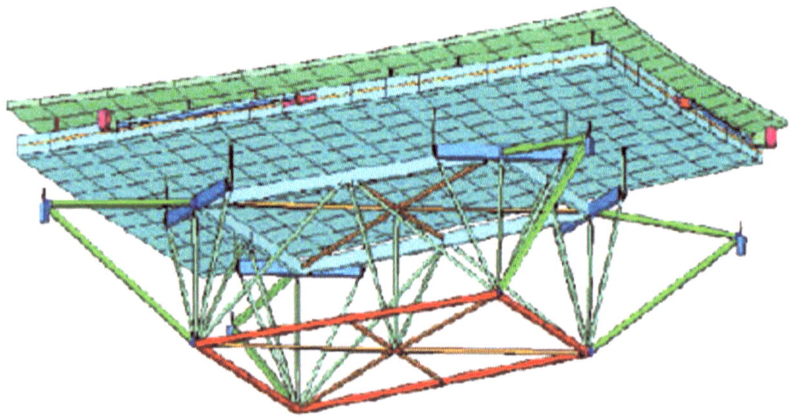

Fig. 7.18. The panel unit of the LMT. The red-green subframe is connected to the BUS by 4 actuators (shown in blue on the green spars). The cyan rectangle with 4 blue beams is the isostatic structure providing axial support of the panel baseplate (light blue). The green top layer is the reflecting surface.

We conclude the discussion of the LMT with a few remarks on the most important part of the telescope: the reflector surface. As with the MRT, it was decided to subdivide the surface in a relatively small number of panel units, which, in this case, will be supported on their four corners by the motorised adjusters. The reflecting skin of the units will be pre-adjusted in the shop to the prescribed paraboloid, so that on the telescope only the larger units need to be measured and adjusted. The units are about 5 x 2 m^2, suitable for transportation and handling, leading to a total of 180 units and 720 adjusters. The error allowance of the panel units is 25 μm. This cannot be achieved by only a "hard" support of the unit corners; the center would sag by about 300 μm. This is solved by the introduction of an "isostatic" intermediate support, made of stainless steel, between the surface skin and the four-point support, as illustrated in Fig. 7.18. The intermediate support functions similar to the "whiffle tree" support used in optical telescopes, notably the 10 m diameter Keck telescopes on Hawaii (Nelson et al., 1983). The baseframe of aluminium is supported on 8 points by the isostatic structure and shows a deformation under gravity of only 10 μm. This baseplate will carry the surface, for which several options are available. One considered is a continuous CFRP sheet supported on some 200 adjusters and pre-adjusted in the shop. Another possibility is the use of a number (8) of separate surface plates. For the inner 3 of the 5 panel unit rings, the surface will indeed be composed of 8 subpanels per unit, fabricated in a new technology, developed by the company Media Lario of Italy for the European ALMA prototype antenna. Because historically this development came before its application to the LMT, we summarise this technology below in the section on the ALMA antennas. These panels can be seen in Fig. 7.19, where the inner rings are being installed.

Fig. 7.19. View of the LMT backup structure with the second and third panel ring being installed (Sep. 2006). The approximately 5x2 m^2 panels each carry 8 surface plates. Note the yellow dust settled onto the surface at the 4600 m high site. (E. Mendez, INAOE)

Once operational the LMT will provide exciting new observing capabilities. At the time of writing the mechanical assembly of the telescope is in an advanced state (see Fig. 7.20). The project is plagued by cost overruns and the date of full operation is uncertain. Once completed the telescope will add a collecting area of about 2000 m^2 for short mm-wavelength astronomy, a significant increase in the world's supply.

But, as is the case with all single, full aperture telescopes, the angular resolution, even at the level of 5-10", is still about an order of magnitude worse than that of optical telescopes. A significantly higher angular resolution can only be achieved by the application of interferometry and aperture synthesis. In the last section of this chapter we shall discuss some of the features of the antennas of a high resolution array for submillimeter wavelengths as short as 0.3 mm.

Fig. 7.20. The 50-m diameter LMT under construction at an altitude of 4600 m on Cerro la Negra, central Mexico in summer 2006. (E. Mendez, INAOE)

■ 7.8. The ALMA Prototype Antennas

For more than 15 years astronomers in the USA, Japan and Europe have been making plans for a large millimeter array, placed at a superb site. In 2000 these plans were merged into the ALMA project, which is joint project of North America (USA and Canada) and Europe (ESO, representing 12 European countries) under participation of Japan and Taiwan. The project comprises 54 antennas of 12 m diameter, suitable for observations at 1 THz (0.3 mm wavelength) and equipped with receivers in the entire frequency range from 75 - 950 GHz. In addition, there will be a "compact array" consisting of 12 antennas of 7 m diameter. The instrument is located at 5000 m altitude in the Atacama Desert of Northern Chile, hence the name Atacama Large Millimeter Array (ALMA). Construction is under way and the full array should be completed by 2012.

In the original independent concepts for a millimeter array differences in scientific emphasis among the groups led to the choice of quite different antenna diameters. For the Americans the detailed mapping of extended objects was of highest priority. This led to the proposal to build an array of 40 antennas of 8 m diameter. In Europe the emphasis was on the observation of weak extragalactic objects of relatively small angular size. Hence there was a need for high sensitivity and the choice fell on dishes of 15 m diameter. The Japanese considered a diameter of 10 m as suitable for their purpose. Eventually, a compromise was reached to select a diameter of 12 m with the requirement that the antennas should operate well to a shortest wavelength of 0.3 mm.

Because the project plan contained up to 64 of these antennas, it was decided to develop a prototype first. Both the US and European partner placed a contract for a prototype antenna with two different companies. The specifications are stringent: surface rms 25 μm with a goal of 20 μm, all sky absolute pointing 2" and "local" pointing and tracking stability 0.6" in 9 m/s wind and full thermal loading by the Sun at 5000 m altitude. There are two new unusual specifications which follow directly from the interferometric mode of operation. Any differential pathlength variation in the antenna will cause a change in the recorded phase of the interferogram and be indistinguishable from such a change caused by the brightness distribution of the observed object. Thus the antennas are required to limit pathlength variations in the structure to less than 15 μm over a time of 15-30 minutes. Finally, a widely used mode of observation will consist of switching at short time intervals, in the order of 10 seconds, between the main object of investigation and a calibration radio source, a short distance on the sky away. To enable this to be done effectively, the antennas must be able to switch between two positions on the sky 1.5 degree apart within 1.5 second of time. This puts high requirements on the drive and control system.

The antennas will moreover be moved between some 200 baseline stations spread across the site with maximum distances of 14 km. Thus a low weight is preferred, but the antenna must be able to withstand the extreme conditions on the site, which includes very strong winds and occasional snow or icing. The US company Vertex-RSI, together with their German subsidiary Vertex Antennentechnik (the group which designed the HHT, before it was sold by Krupp to Vertex), produced an antenna under contract to the US National Radio Astronomy Observatory (NRAO), while a consortium of the European companies Alcatel and European Industrial Engineering (EIE), here called AEC, delivered an antenna to the European Southern Observatory (ESO). Both antennas were erected at the site of the NRAO Very Large Array in New Mexico (US) and subjected to an extensive program of technical evaluation. The author participated in this program and in the following pages the design features of both antennas are summarised and some aspects of the measurement methods and equipment are presented.

On the basis of the performance of the few existing submillimeter antennas, both design groups decided to make extensive use of CFRP. It was expected that this would avoid the need to incorporate an active surface. It was also felt that the demanding pointing accuracy would necessitate the use of metrology (Flexible Body Compensation) to correct for deformations unseen by the axis position encoders. Finally, an economic solution had to be found for the mass production of the surface panels at an accuracy better than 10 μm for a total area of almost 8000 m^2. These all formed significant challenges to the design engineers.

7.8.1. The VertexRSI design

The VertexRSI antenna is shown in Fig. 7.21. The mount is traditional, made from steel, covered with thermal insulation on the outside. The base provides a three-point

Fig. 7.21. Prototype 12 m diameter antenna for ALMA, designed and built by VertexRSI. Note the curved quadripod to minimise the spherical wave aperture blocking. (N.Emerson, NRAO/AUI)

connection to the foundation, from which it can quickly be released for transportation of the antenna to another baseline station. The fork of the alidade runs on a roller bearing of 2.5 m diameter. The elevation structure rotates on two bearings in the top of the yoke, providing a large receiver room directly behind the BUS. A receiver box of 1 m^3 can be accommodated in this room. The elevation structure is made of steel, thermally insulated on the outside. The connection from the receiver cabin to the BUS is realised in invar steel to absorb the large difference in thermal expansion between the steel of the receiver cabin and the CFRP of the BUS.

The BUS is a box structure made from composite plates consisting of CFRP skins bonded to aluminium honeycomb cores. There are 24 sectors providing support for the surface panels. Vertex has not used their proven space-frame solution, pioneered with the HHT, thus avoiding the expensive fabrication of the nodes in CFRP. Although this was not exercised for one prototype, the box structure lends itself quite well to the use of assembly templates in the series fabrication of the antennas. Consequently the choice of the box structure undoubtedly is the cheaper solution. These structures can be analysed well by modern finite element methods and the deviation from homology in this case is only 5 μm rms. As we have seen before, apart from the lower weight, the CFRP BUS exhibits a very good thermal behaviour.

As clearly visible in Fig. 7.21, the quadripod support of the secondary reflector is strongly curved. This tends to decrease the spherical-wave blocking (see Sec.4.5). It was first used in the satellite ground station at Raisting, Germany in 1964. In the ALMA design, the Vertex group has made the quadripod part of the load-bearing structure, which results in a higher stiffness of the BUS-quadripod combination. The geometrical blocking percentage lies just under 3 percent, as specified.

The reflector surface is composed of 8 rings of aluminium panels, for a total of 264, each about 0.5 m^2 in area. The surface is machined to an rms accuracy of 6 μm on an accurate milling machine. The relatively small size of the panels is caused by the limited area coverage of the milling machine. Because the observation of the Sun for sizeable periods of time is a requirement, the reflector surface must be specially treated to avoid concentration of solar heat on the subreflector and the secondary focus. Vertex tried several ways of scattering the solar visible radiation by the slight grooves of the cutting edge, but they proved unsatisfactory because of the "grating-like" reflection patterns. The final solution is chemical etching of the surface, which gives it a slight roughness at the 1 μm level, which scatters the visible light sufficiently.

The drives are pinion and gear-rack systems powered by dual dc-motors in anti-backlash configuration. To meet the very high pointing requirements Vertex includes a metrology system in the mount to determine the pointing errors which are not sensed by the angle encoders. It consists of an independently supported CFRP structure in the forkarms which is connected through displacement sensors to the elevation bearings. Together with a set of temperature sensors in the steel part of the antenna these should help to correct for pointing changes due to wind and temperature variations.

The total weight of the VertexRSI prototype ALMA antenna is 108 tonnes, equally divided over the azimuth and elevation sections.

7.8.2. The Alcatel-EIE-Consortium (AEC) design

The AEC antenna is shown in Fig. 7.22. The mount of this antenna is also realised in steel. The base is supported on the foundation by six flanges, which absorb the difference in thermal expansion between the base and the concrete foundation. This is not as specified for the final array and the production antennas will have a three point support. The fork runs on a 2.6 m diameter roller bearing and the elevation structure rotates on two elevation bearings on top of the fork. In this design the entire moving part of the elevation structure is made of CFRP, including the receiver cabin. This minimises both the weight and the thermal expansion of this structure. The BUS is a box structure consisting of 16 sections. The members are full CFRP plates bonded and bolted together to form the BUS. The computed homology deviation is 6 μm. The quadripod is a thin, straight structure and the geometrical blocking is also just under 3 percent.

Fig. 7.22. Prototype 12 m antenna for ALMA designed and built by Actatel - European Industrial Engineering Consortium. This antenna employs direct drives with linear motors. The magnet arc of the elevation drive is well visible. (N. Emerson, NRAO/AUI)

The reflector is composed of 120 panels with an average area of about 1 m^2, placed on the BUS in 5 concentric rings. A new panel technology has been developed by the company Media Lario of Italy. It is an application of electroformed, replicated surface skins, developed for the XMM (X-ray) satellite of the European Space Agency. In our application mechanically figured steel molds are used to deposit nickel skins of about 1.5 mm thickness in a chemical electroform procedure. The form and surface characteristics of the replicated skins match the mold extremely precisely. The skins are then bonded to a core of aluminium honeycomb to form a stiff, lightweight panel. The panel accuracy is essentially determined by that of the mold and for the ALMA antenna the panels were manufactured with a surface rms of about 8 μm. As mentioned above, such panels are also used on the LMT in Mexico. An effective scattering of solar visible light (heat) is assured by the roughness at the 1 μm scale of the steel molds. A slight disadvantage of the nickel surface is its relatively large absorptivity for heat. This is counteracted by a 200 nm thin layer of rhodium deposited onto the final front surface of the panel. An additional advantage of this is the excellent protection against environmental influences. The replication technique is especially suitable for mass productions, as will be needed for the fabrication of the series of 25 or more ALMA production antennas.

The drive system of the AEC antennas is unusual for a radio telescope. It employs direct drives with linear motors, hence it totally avoids gearboxes. The azimuth drive is placed in a circle near the large roller bearing. The elevation drive is located centrally along the outside of the receiver cabin. There is essentially no backlash with this drive system and the drive and control system show an especially good performance in pointing and tracking stability under wind influence. Nevertheless, to meet the stringent pointing requirements AEC incorporates a metrology system to measure the pointing errors which are not detected by the axis encoders. It consists of a laser-based system in the forkarms to measure the shifts of the elevation bearings, together with a set of temperature sensors in the steel section of the antenna. No such sensors are needed in the elevation section, because it is entirely realised in CFRP.

The weight of the AEC prototype ALMA antenna is 80 tonnes, of which 65 percent is taken by the azimuth structure and 35 percent by the elevation section.

7.8.3. Performance evaluation of the ALMA prototype antennas

The two ALMA prototype antennas were placed next to each other and subjected to an intensive evaluation program. In this section we summarise some of the methods applied for this evaluation. The results of the program have been published; the performance of the antennas is summarised by Mangum et al. (2006), while several papers have been written with descriptions of the measurement equipment and evaluation methods (Greve and Mangum (2006), Snel et al.(2007) and Baars et al.(2006)).

The evaluation faced several challenges. First of all, no antenna had been built with this combination of size and accuracy. Secondly, the site of the VLA in New Mexico does not offer the atmospheric conditions necessary for observations at submillimeter wavelengths. Thus the usual test procedures, as described in this book, were of limited use and we were forced to devise other measurement methods and the accompanying instruments to perform a satisfactory performance evaluation.

In winter the VLA site will occasionally allow observation near 1 mm wavelength. Considering that the reflector accuracy should be better than 25 μm rms, it was out of the question to use aperture efficiency measurements, as described in Sec.5.4., to check this quantity. We used near-field radio holography at 3 mm wavelength to measure and set the surfaces at a single elevation angle of 9 degrees. The discussion of such a measurement program has been presented in Sec.6.1 with examples taken from the work on the ALMA antennas. The surfaces were set to an accuracy of about 16 μm with an estimated uncertainly of 5 μm. For details see Sec.6.1 and Baars et al. (2006).

The antennas were equipped with a receiver for 3 and 1.2 mm wavelength. Given the deficiencies of the atmosphere at the site, combined with the sparsity of sufficiently strong radio sources at these wavelengths, it would essentially be impossible to determine a good pointing model for the antennas. To ameliorate the situation, each antenna was equipped with an optical telescope and CCD-camera, mounted in the stiff BUS looking at the sky through a hole in the reflector. Using the abundantly available stars it is now possible to gather data for a good pointing model of the antenna up to the BUS structure. We still need radio observations to determine how the radio beam axis deviates from the optical line of sight as function of elevation angle and under varying conditions as changing temperature. However, the optical telescope is an invaluable help in determining a starting pointing model and studying pointing stability of the mount and BUS under wind influence. Once the optical pointing model was available, the millimeter receivers were used to determine the offsets to the radio pointing by the use of a limited number of sources.

By placing the half-power point of the antenna beam on the radio source, small variations in the pointing can be detected by the relatively large change in output signal. This method was used to obtain a reasonably good impression of the tracking stability under wind. Further information on the dynamical behaviour of the antenna under wind influence was obtained from a set of 10 accelerometers which were placed at the rim of the BUS, in the apex near the secondary reflector and on the stiff central flange for the receiver box, behind the vertex of the primary reflector. By double integration of the measured accelerometer signal, data could be obtained about the actual positional changes of the sensors. This enabled us to measure both pointing variations of the order of 0.1 arcsecond and path length changes of a few micrometers in realistic situations under actual wind influence. This diagnostic tool turned out to be extremely useful for the evaluation of the antennas and could certainly find application in other cases. The details of the method can be found in Snel et al. (2007).

The unusual requirement of path length stability also forced us to devise a measurement method for this quantity. This used essentially optical distance and displacement measurements carried out with lasers, quadrant detectors and an API 5-degrees of freedom laser-interferometer. Several critical dimensions in the antenna were checked with this equipment as function of elevation angle and with varying temperature. The measurement accuracy was a few micrometers and it enabled us to demonstrate the adherence of the antennas to the specification (Greve and Mangum, 2006). Some path length changes could also be measured with the aid of radio sources. For instance, the variation of focal length, both as function of elevation angle and temperature was shown in Sec. 5.5.

With the aid of these test methods, the mediocre quality of the site notwithstanding, it has been possible to demonstrate that both ALMA prototype antennas satisfy the stringent specifications. An important factor is the extensive use of CFRP in the antenna structure. It is the first instance in which a box-type backup structure provides a reflector support of this high accuracy and stability. Beyond this, there are no entirely new design aspects or technologies employed. The antennas are however very well designed and the fabrication properly executed. The new additions of the mechanical or optical metrology systems did not perform to specification during the evaluation phase. It is all the more remarkable that the antennas perform close to their specifications without these systems activated. Both ALMA prototype antennas are by all consideration very good pieces of engineering.

The original idea behind acquiring two prototypes was of course to increase the chance for a successful design satisfying the very high demands and also to select the best or cheapest on the basis of the evaluation. Eventually several political, managerial and financial circumstances lead to the decision to order half of the antennas from each of the two companies and let each of them deliver antennas of their own design. It is hard to guess what would have happened if one of the prototypes had been much better, much cheaper, or both than the other one. But in the current course of the ALMA Project it is fortunate that both antennas comply with the specifications.

■ 7.9. Conclusion

The aim of the short descriptions of these radio telescopes has been to highlight some of the original design aspects of major antennas over the last 40 years. We have now entered the era in which extensive metrology and active control of the reflector surface and pointing fluctuations are necessary to maintain performance under operational conditions. The GBT and LMT are examples of this approach. With access to the superb sites in the Atacama Desert and the South Pole, astronomers are proposing to build submillimeter telescopes with a highest frequency well above 1 THz, and larger than existing ones. One example is the Cornell Caltech-Atacama-Telescope (CCAT). The proposal entails a 25 m diameter reflector of 10 μm rms surface accuracy and a pointing accuracy of 0.35 arcsecond. On the other side of the spectrum, the Square Kilometer Array (SKA), operating in the frequency band 100 MHz - 25 GHz might be realised by some 6000 dishes of approximately 15 m diameter. Here the challenge is not so much surface and pointing accuracy as low production cost. A similar approach of an array of relatively small dishes is being considered for the next generation of NASA's deep-space tracking stations to replace the aging 70 m antennas.

In the meantime, existing telescopes are adopting active elements in order to improve the performance. An example is the new "deformable" subreflector for the Effelsberg telescope to correct for the large-scale deformations of the primary reflector.

Finally, the optical astronomers are designing telescopes in the 25 - 100 m class. Some of these look remarkably similar to radio telescopes, but of course they require a glass reflector with an accuracy and smoothness measured in nanometers instead of micrometers. However, building and commissioning these will require entirely new methods and materials, a possible subject for another book by a different author.

References

Altmann, H., Die Stahlkonstruktion des 100-m Radioteleskops in Effelsberg, *Der Stahlbau* **41**, 321 ff and 360 ff, 1972.

Baars, J.W.M., *Technology of large Radio Telescopes for millimeter and submillimeter wavelengths*, Infrared and Millimeter Waves, **9**, 241-281, K.J. Button, Ed., New York, Academic Press, 1983.

Baars, J.W.M. and B.G. Hooghoudt, The Synthesis Radio Telescope at Westerbork. General Lay-out and Mechanical Aspects, *Astron. Astrophys.* **31**, 323-331, 1974.

Baars, J.W.M., B.G. Hooghoudt, P.G. Mezger and M.J. de Jonge, The IRAM 30-m millimeter radio telescope on Pico Veleta, Spain. *Astron. Astrophys.* **175**, 319-326, 1987 [Gold, 56].

Baars, J.W.M. A. Greve, H. Hein, D. Morris, J. Penalver and C. Thum, Design parameters and measured performance of the IRAM 30-m millimeter radio telescope, *Proc. IEEE* **82**, 687-696, 1994.

Baars, J.W.M, R.N. Martin, J.G. Mangum, J.P. McMullin, W.L. Peters, The Heinrich Hertz Telescope and the Submillimeter Telescope Observatory, *Publ. Astron. Soc. Pacific* **111**, 627-646, 1999.

Baars, J.W.M., J.G. Mangum, R. Lucas and J.A. Lopez-Perez, Near-Field Radio Holography of Large Reflector Antennas, *IEEE Antennas and Propagation Magazine* **48**, 2006.

Bowen, E.G. and H.C. Minnett, The Australian 210-foot Radio Telescope, *Proc. Inst. Radio Eng. Australia* **24**, 98-105, 1963.

Brandt, P. and H. Gatzlaff, Das Design des 30-m Millimeterwellen-Radioteleskops, *Krupp Forschungsberichte* **39**, 111-124, 1981.

Christiansen, W.N., W.C. Erickson and J.A. Högbom, The Benelux Cross Antenna Project, *Proc. Inst. Radio Engineers Australia* **24,** 219 - 225, 1963.

Eschenauer, H., H. Gatzlaff and H.W. Kiedrowski, Entwicklung und Optimierung hochgenauer Paneeltragstrukturen, *Tech. Mitt. Krupp Forschungs Ber.* **38**, 43-57, 1980.

Geldmacher, E., 100-m Radioteleskop für das Max-Planck-Institut für Radioastronomie Bonn, *Tech. Mitteilungen Krupp* **28**, 187-197, 1970.

Gordon, M.A., *Recollections of "Tucson Operations"* , New York, Springer, 2005.

Greve, A., J.W.M. Baars, J. Peñalver, and B. LeFloch, Near focus active optics: an inexpensive method to improve millimeter wavelength radio telescopes, *Radio Sci.* **31**, 1053-1066, 1996.

Greve, A., M. Bremer, J. Peñalver, P. Raffin and D. Morris, Improvement of the IRAM 30-m Telescope from Temperature Meausurements and Finite-Element Calculations, *IEEE Trans. Antennas and Propag.* **AP-53**, 851-860, 2005.

Greve, A. and J.G. Mangum, Mechanical Measurements of the ALMA Prototype Antennas, *IEEE Antennas and Propagation Magazine* **49**, 2007.

Guilloteau, S., J. Delannoy, D. Downes, A. Greve, M. Guelin, R. Lucas, D. Morris, S.J.E. Radford, J. Wink, J Cernicharo, T. Forveille, S. Garcia-Burillo, R. Neri, J. Blondel, A. Perrigourad, D. Plathner, and M. Torres, The IRAM interferometer on Plateau de Bure, *Astron. Astrophys.* **262**, 624-633, 1992.

Hachenberg, O., Studien zur Konstruktion des 100-m Teleskops, *Beiträge zur Radioastronomie (MPIfR)* **1**, 31-61, 1968.

Hachenberg. O., B.H. Grahl and R. Wielebinski, The 100-meter Radio Telescope at Effelsberg. *Proc. IEEE* **61**, 1288-1295, 1973 [Gold, 32].

Hooghoudt, B.G., The Benelux Cross Antenna Project, *Annals. New York Acad. Sci.* **116**, 13-24, 1964.

Jacobs, E. and H.E. King, A 2.8-minute beamwidth, millimeter-wave antenna - measurements and evaluation, *IEEE Int. Conv. Rec.* **13**, Pt.5, 92-100, 1965.

Kärcher, H.J., Enhanced pointing of telescopes by smart structure concepts based on modal observers, *SPIE Proceedings* **3668**, 998-1009, 1999.

Karcher, H.J., Telescopes as mechatronic systems, *IEEE Antennas and Propagation Magazine* **48**, No.2, 17-37, 2006.

Kärcher, H.J. and J.W.M. Baars, The design of the Large Millimeter Telescope / Gran Telescopio Milimétrico (LMT/GTM), *SPIE Proceedings,* **4015**, 155-168, 2000.

Leighton, R.B., A 10-meter telescope for millimeter and sub-millimeter astronomy, Tech. Rep California Inst. of Technology, May 1978.

Mäder, H.F., K.-H. Stenvers, H. Gatzlaff and H.-F. Wilms, The 10-m Submillimeter-wave Telescope, *Technische Mitteilungen Krupp*, No.1, 27-42, 1990.

Mangum, J.G., Baars, J.W.M., Greve, A., Lucas, R., Snel, R., Wallace, P.T., & Holdaway, M. Evaluation of the ALMA Prototype Antennas, *Publ. Ason. Soc. Pacific* **118**, 1257-1301, 2006.

Mezger, P.M., *Principle considerations of radioastronomical observations at very high frequencies*, NRAO unnumbered Internal Report, March 1964.

Nelson J.E., M.P. Budiansky, G. Gabor and T.S. Mast, The segmented-mirror control system prototype for the ten meter telescope, *Proc. SPIE* **444**, 274-286, 1983.

Ryle, M., The new Cambridge Telescope, *Nature* **194**, 517-518, 1962.

Snel, R.C., J.G. Mangum and J.W.M. Baars, Study of the Dynamics of Large Reflector Antennas with Accelerometers, *IEEE Antennas and Propagation Magazine* **49**, 2007.

Stenvers, K.-H. and H.-F. Wilms, Entwicklung und Optimierung eines tragenden CFK-Raumfachwerks für astronomische Teleskope, *VDI-Berichte* **734**, 75-92, 1989.

Tolbert, C.W., A.W. Straiton and L.C. Krause, A 16-foot diameter millimeter wavelength antenna system, its characteristics and its applications, *IEEE Trans. Ant. Prop.* **AP-13**, 225-229, 1965.

von Hoerner, S., Design of large steerable antennas, *Astron. J.* **72**, 35-47, 1967a [Gold, 64].

von Hoerner, S., Homologous deformations of tiltable telescopes, *J. Struct. Division Proc. Am. Soc. Engrs.* **93**, 461 -485, 1967b.

von Hoerner, S. and W.Y. Wong, Gravitational deformation and astigmatrism of tiltable radio telescopes, *IEEE trans. Antennas Propagation* **AP-23**, 689-695, 1975.

Wilson, R.W., K.B. Jeffers and A.A. Penzias, Carbon monoxide in the Orion Nebula, *Astophys. J.* **161**, L43, 1970.

Name Index

Reference pages in *italics*

Subject Index

MIX
Papier aus verantwortungsvollen Quellen
Paper from responsible sources
FSC® C105338

If you have any concerns about our products,
you can contact us on
ProductSafety@springernature.com

In case Publisher is established outside the EU,
the EU authorized representative is:
Springer Nature Customer Service Center GmbH
Europaplatz 3, 69115 Heidelberg, Germany

Printed by Libri Plureos GmbH
in Hamburg, Germany